A Biosystems Approach to Industrial Patient Monitoring and Diagnostic Devices

A Biosystems Approach to Industrial Patient Monitoring and Diagnostic Devices
Gail Baura

ISBN: 978-3-031-00497-1   paperback

ISBN: 978-3-031-01625-7   ebook

DOI 10.1007/978-3-031-01625-7

A Publication in the Springer series
SYNTHESIS LECTURES ON BIOMEDICAL ENGINEERING #12

Lecture #12
Series Editor: John D. Enderle, University of Connecticut

First Edition
10 9 8 7 6 5 4 3 2 1

A Biosystems Approach to Industrial Patient Monitoring and Diagnostic Devices

Gail Baura
Keck Graduate Institute
Clarmeont California

SYNTHESIS LECTURES ON BIOMEDICAL ENGINEERING #12

ABSTRACT

A medical device is an apparatus that uses engineering and scientific principles to interface to physiology and diagnose or treat a disease. In this Lecture, we specifically consider those medical devices that are computer based, and are therefore referred to as medical instruments. Further, the medical instruments we discuss are those that incorporate system theory into their designs. We divide these types of instruments into those that provide continuous observation and those that provide a single snapshot of health information. These instruments are termed patient monitoring devices and diagnostic devices, respectively. Within this Lecture, we highlight some of the common system theory techniques that are part of the toolkit of medical device engineers in industry. These techniques include the pseudorandom binary sequence, adaptive filtering, wavelet transforms, the autoregressive moving average model with exogenous input, artificial neural networks, fuzzy models, and fuzzy control. Because the clinical usage requirements for patient monitoring and diagnostic devices are so high, system theory is the preferred substitute for heuristic, empirical processing during noise artifact minimization and classification.

KEYWORDS

System theory, Machine intelligence, Patient monitoring, In vitro diagnostics, Pseudorandom binary sequence, Adaptive filtering, Wavelet transforms, ARMAX model, Artificial neural networks, Fuzzy model, Fuzzy control

To Larry Spiro, my bon vivant, without whose infinite patience and love this book could not have been written.

Contents

Preface

While system theory applied to patient monitoring and diagnostic devices continues to be an obscure specialty, it remains a specialty that has improved the diagnosis and assessment of interventions in countless patients. This text may be used as a supplement to a traditional medical instrumentation course in biomedical engineering.

I would like to thank Dr. John Enderle and Joel Claypool for encouraging me to write an undergraduate summary of this specialty, and Dr. Sandy Ng for reviewing the manuscript. I would also like to thank my Ph.D. advisors, Drs. David Foster and Dan Porte, Jr., for their continuing support.

My husband, Larry, continues to understand my obsessive endeavors and patiently wait for their completion. In our twenty years together, he has opened my mind up to so much more than system theory. And I continue to apologize that my first book wasn't my last book.

I welcome comments to this text at www.gailbaura.com.

Gail D. Baura
Claremont, 2007

CHAPTER 1

Medical Devices

In 1976, as part of the Medical Device Amendments to the Federal Food, Drug, and Cosmetic Act, the term "medical device" was defined as:

"an instrument, apparatus, implement, machine, contrivance, implant, in vitro reagent, or other similar or related article, including a component part, or accessory which is:

- recognized in the official National Formulary, or the United States Pharmacopoeia, or any supplement to them,
- intended for use in the diagnosis of disease or other conditions, or in the cure, mitigation, treatment, or prevention of disease, in man or other animals, or
- intended to affect the structure or any function of the body of man or other animals, and which does not achieve any of its primary intended purposes through chemical action within or on the body of man or other animals and which is not dependent upon being metabolized for the achievement of any of its primary intended purposes" (United States Code 1976).

Basically, a medical device is an apparatus that uses engineering and scientific principles to interface to physiology and diagnose or treat a disease. You will note that an "in vitro reagent", which is part of an in vitro diagnostic, is specifically listed. For this reason, in vitro diagnostics are classified by the Food and Drug Administration (FDA) as a type of medical device. Further, since this definition distinguishes a device as an apparatus that affects physiologic structure or function *without chemical action*, it serves to differentiate a device from a drug or biologic.

Medical devices are classified on the basis of patient risk. They span the classification range from low risk, or Class I, to moderate risk, or Class II, and significant risk, or Class III. An example of a Class I device is a tongue depressor. Examples of Class II and Class III devices are a noninvasive patient monitor, such as a cardiograph that measures heart activity, and an implantable orthopedic prosthesis, respectively. In general, the classification of a device dictates the rigor with which it is regulated, in terms of FDA's processes for premarket approval and postmarket surveillance of device complications.

1.1 MEDICAL DEVICE INDUSTRY

In the United States, there are approximately 6000 medical device companies, covering 50 clinical specialties. Only about 100 of these companies produce annual revenues over $100 million. Approximately 72% of these manufacturers employ fewer than 50 people each (Marwick 2000). According to a 2003 survey by the United States Department of Labor, 15,790 engineers were employed in this industry in 2003 (United States Department of Labor 2006).

Overall, these companies have been very successful. Three years after regulation began, the U.S. medical device industry shipped $9.8 billion of goods in 1979. By 2004, these shipments had grown by approximately an order of magnitude to $93.8 billion. As shown in Figs. 1.1 and 1.2, the mix of devices has not changed dramatically over 25 years. In both 1979 and 2004, surgical/medical instruments and surgical appliances/supplies accounted for the largest

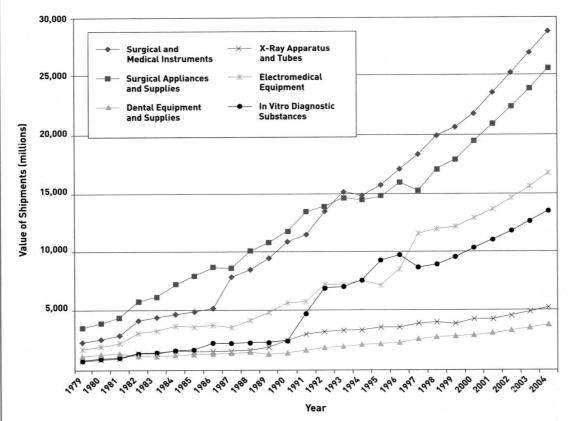

FIGURE 1.1: Comparison of U.S. medical device sector percentage shipments in 1979 and 2004. Reprinted with permission from *Medical Device & Diagnostic Industry*, "The Medical Device Market: Up, Up, and Away," August 2004. Copyright © 2004 Canon Communications LLC.

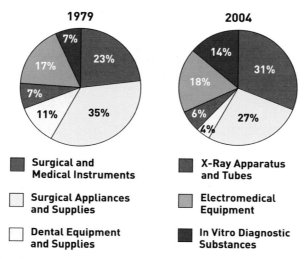

FIGURE 1.2: Growth of medical device shipments by device sector from 1979 to 2004. Reprinted with permission from *Medical Device & Diagnostic Industry*, "The Medical Device Market: Up, Up, and Away," August 2004. Copyright © 2004 Canon Communications LLC.

shipment percentages. Similarly, in both years, x-ray apparatus/tubes accounted for one of the smallest shipment percentages (Conroy 2004).

Currently, worldwide heart rhythm management and coronary artery stent sales are $10.6 and $6.5 billion markets, respectively. Similarly, obesity management, knee replacement, and hip replacement sales are $190 million, $4.8 billion, and $4.3 billion markets, respectively. Device companies are forecasted to continue to grow because of the aging population and innovation. According to the Wall Street firm Lazard Capital Markets, the number of Americans aged over 60 is projected to increase by at least 30% over the next eight years. On average, they will spend more on healthcare between the ages of 60 and 70 than they did in all their previous decades combined. Many successful devices emerged in response to the limitations of drugs. For example, pacemakers and defibrillators correct heartbeat abnormalities that cannot be treated chemically. Devices can be designed with electronic components that link them to computers and communications networks, allowing data from patients with implants to be monitored remotely (Feder 2006).

1.2   MEDICAL INSTRUMENTATION

Medical instrumentation manufacturers are a subset of medical device manufacturers, distinguished by manufacture of devices that incorporate embedded or PC-based systems for use in diagnostic and patient monitoring applications (Baura 2002). Pacemakers use embedded systems for detection of heartbeats and stimulation after abnormal beat detection. Alternatively,

Medical Graphics Corporation manufactures a spirometer for measurement of lung volumes. The spirometer consists of Windows-based software, a PC card, and a pneumotach for conversion of airflow to a differential pressure (Medical Graphics Corporation 2001). Within medical instrumentation, we can further subdivide manufacturers into those who manufacture patient monitoring equipment, and those who manufacture diagnostic equipment.

1.3   PATIENT MONITORING DEVICES

Patient monitoring refers to the continuous observation of repeating events of physiologic function to guide therapy or to monitor the effectiveness of interventions. Historically, these medical instruments have been most widely deployed in the intensive care units (ICUs) and operating rooms of hospitals, for care of the most critically ill patients. Some commonly processed signals are the electrocardiogram, intraarterial blood pressure, and arterial saturation of oxygen, which is also known as pulse oximetry. Over time, patient monitoring has moved from invasive to noninvasive measurements, which decrease patient risk.

Traditionally within ICUs, Hewlett Packard, Marquette, and Spacelabs were the dominant manufacturers of racks of modules that provided patient monitoring. With the onset of mergers and acquisitions (M&As) in the 1990s, Hewlett Packard spun off Agilent, which was then purchased by Philips. Marquette was purchased by General Electric. Through a series of M&As, Spacelabs became part of OSI Systems. Today, these racks have been compressed into smaller integrated systems, such as the Philips Intellevue MP70 patient monitor (Fig. 1.3).

1.3.1   Nellcor Disposable Model

Along with the trend of noninvasive measurement, patient monitoring tends to incorporate disposable sensors. This marketing strategy was first implemented by Nellcor in the mid 1980s. Although Biox Technology was responsible for the enabling technology of a calibration curve for pulse oximetry, Nellcor captured a larger market share by becoming the first manufacturer of disposable pulse oximetry sensors. Nellcor marketed this feature to address sterility concerns during the height of the AIDS crisis and cost containment. Because a different calibration curve is needed for each specific sensor–monitor combination, Nellcor argued that, with standardization of Nellcor pulse oximeters through the hospital, the same disposable sensor could be moved between hospital units and decrease costs.

Biox was later acquired by Ohmeda. In 1992, Ohmeda sued Nellcor to invalidate four sensor patents. In 1995, the U.S. District Court of Delaware upheld the Nellcor patents, including key patent U.S. 4,621,643 (New and Corenman 1986). As a result, if Ohmeda sold their sensors for use in a non-Ohmeda oximeter, Ohmeda would be infringing on Nellcors' patents (Health Industry Today 1995). By 1997, U.S. disposable sensor sales, excluding sales in federal hospitals and nursing homes, reached $197.5 million. Nellcor accounted for 88% of this

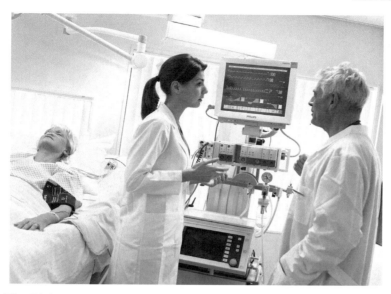

FIGURE 1.3: The Philips Intellevue MP70 patient monitor. © 2007 Koninklijke Philips Electronics N.V. All rights reserved. Reproduction in whole or part is prohibited without prior written consent of the copyright owner.

market. In contrast, U.S. sales of oximeters were only $32 million (IMS Health 1997, Baura 2002).

1.4 DIAGNOSTIC DEVICES

Medical diagnostics refers to discrete testing conducted to provide critical health care information for risk assessment, early diagnosis, treatment, or disease management. In the 17th century, clinicians diagnosed diabetes through the level of sweetness in urine samples. By the end of the 18th century, when the first hospital laboratory was established in Britain, diagnostics started to become recognized as a standard and indispensable part of healthcare. Current in vivo, or within the body, diagnostic tests include the computed tomography scan, magnetic resonance imaging, and blood pressure screening. Current in vitro, or from an external body specimen, diagnostics (IVD) tests include cholesterol, Papanicolaou (Pap) smear, and conventional glucose monitoring tests (Baura 2006).

General Electric, Siemens, and Philips dominate in vivo diagnostics because of their well-established portfolios of imaging systems. Roche Diagnostics, Abbott Diagnostics, and Ortho Clinical Diagnostics are the largest IVD manufacturers. Roche Diagnostics includes its 1998 acquisition—Boehringer Mannheim. Roche's Cobas 6000 platform, which automates clinical chemistry and immunoassay testing, is shown in Fig. 1.4. Abbott Diagnostics is part of

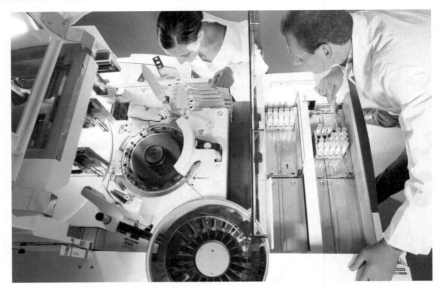

FIGURE 1.4: Roche Cobas 6000 Chemistry/Immunochemistry Analyzer. Courtesy of F. Hoffmann-LaRoche, Basel, Switzerland.

Abbott Laboratories, which was founded by Chicago physician Wallace Abbott in 1900. Ortho Clinical Diagnostics is the merger of two Johnson & Johnson companies: Ortho Diagnostic Systems and Johnson & Johnson Clinical Diagnostics (IVD Technology 2005).

1.5   CONCLUSION

The medical device industry has been very successful in a regulated environment for over three decades. Devices have addressed clinical needs that range from aging joints to aberrant heartbeats. Moving forward, future devices will incorporate a host of technologies that enable a larger set of device features to be deployed.

REFERENCES

Baura, G. D., *System Theory and Practical Application of Biomedical Signals.* Hoboken, NJ: Wiley-IEEE Press, 2002.

——— "System theory in medical diagnostic devices: an overview," in *28th Annual Int. Conf. of the IEEE Eng. in Medicine and Biology Society*, New York City, 2006, pp. 137–139.

Conroy, S., "The medical device market: up, up, and away," *MD&DI*, pp. 72–73, Aug. 2004.

Feder, B., "Medical devices are hot, which is why Guidant is," *NY Times*, C1, Jan. 21, 2006.

———, "Nellcor Inc. — wins patent infringement suit against Ohmeda Inc.," *Health Industry Today*, Vol. 1, Sep. 1995.

IMS Health, *Hospital Supply Index*. Plymouth Meeting, PA: IMS America, 1997.

Marwick, C., "Device manufacturers consider constraints," *JAMA*, Vol. 283, pp. 1410–1412, 2000.

——, "Top 10 IVD manufacturers," *IVD Technology*, Vol. 14, Jan./Feb. 2005.

Medical Graphics Corporation. http://www.medgraph.com, 2001.

New, Jr., W. and Corenman, J. E., "Calibrated optical oximeter probe," U.S. Patent 4,621,643, November 11, 1986.

United States Code, Medical Device Amendments of 1976. http://www.law.cornell.edu/uscode, 2006.

United States Department of Labor. http://stats.bls.gov/oes/2003/november/naics4_339100.htm#b00-0000, 2006.

CHAPTER 2

System Theory

Before we can discuss the integration of system theory into medical devices, we need to define system theory. In this chapter, we present an extremely abbreviated description of system theory. For a more thorough treatment, please see Baura (2002b).

System theory is the transdisciplinary study of the synthesis and design of systems, and the analysis of their performance. Generally, we represent such a system for discrete time, k, as a group of inputs, $\mathbf{u}(k)$; a group of outputs, $\mathbf{y}(k)$; and a system operator, $\mathbf{H}(k)$ (Fig. 2.1).

Here, $\mathbf{u}(k)$ and $\mathbf{y}(k)$ are vectors for this multiple input, multiple output (MIMO) system. The system operator maps the inputs into the outputs, and is not necessarily linear.

If the system is known to follow superposition and time-invariance, then the system operator can be simplified to be an impulse response, $\mathbf{h}(k)$. Recall that a system with superposition follows:

$$\mathbf{H}[au_1(k) + bu_2(k)] = a\mathbf{H}[u_1(k)] + b\mathbf{H}[u_2(k)], \qquad (2.1)$$

for arbitrary constants a and b. For time-invariance, a delay of the input sequence must cause a corresponding shift in the output sequence. Specifically, for all delays, k_0, an input sequence with values $u_1(k) = u(k - k_0)$ must produce the output sequence with values $y_1(k) = y(k - k_0)$.

2.1 SYSTEM THEORY FOR PHYSIOLOGIC SIGNALS

Typically, acquired physiologic signals cannot be described by this simple MIMO system. Often, an acquired signal includes some form of signal distortion, as the hospital environment is an infinite source of signal distortion. In older monitors, 60 Hz interference from power lines may distort the signal of interest. During surgery, electromagnetic interference is generated by the electrosurgical unit used for cautery. In unanesthetized patients, patient

$$\mathbf{u}(k) \longrightarrow \boxed{\mathbf{H}(k)} \longrightarrow \mathbf{y}(k)$$

FIGURE 2.1: Discrete time system

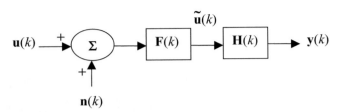

FIGURE 2.2: Discrete time system, accounting for acquired noise. The noise is minimized through filtering

motion is a significant source of distortion. Even respiration and blood pressure may obscure the signal of interest. These and other noise sources make filtering a necessary requirement before further digital signal processing and/or control can be applied to an acquired physiologic signal.

Let us update our system figure to account for acquired noise, $\mathbf{n}(k)$, and subsequent filtering, $\mathbf{F}(k)$ (Fig. 2.2).

You will notice that in this system, noise is assumed to be additive, which may not always be true. Filtering of the combined physiologic and noise signal results in a low noise approximation, $\tilde{u}(k)$, of the true physiologic signal.

Once a low noise approximation of the true physiologic signal is obtained, system identification, or modeling, can be performed to estimate the system operator, $\mathbf{H}(k)$. Typical reasons for modeling the system operator include classification, prediction of future behavior, or insight into underlying physiologic mechanisms.

If it is desired that the outputs be improved, control may be performed on the system (Fig. 2.3).

Please note that the control system operator, $\mathbf{G}(k)$, is not necessarily linear. Also, while the control is illustrated as simple negative feedback, it may be a more complicated nonlinear control such as fuzzy control.

The applications of system theory we have just discussed are illustrated in Fig. 2.4.

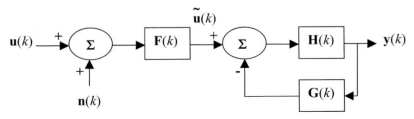

FIGURE 2.3: Discrete time system that utilizes control

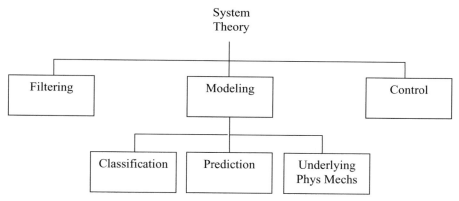

FIGURE 2.4: System theory applications for physiologic signals

2.2   FILTERS

Filtering is an all-purpose term that really refers to noise minimization. If the noise source is restricted to a particular frequency band, frequency-selective filters may be used to minimize the noise. Under certain constraints of a periodic signal, a pseudorandom binary sequence filter may be used, which essentially functions as a bandpass filter that does not require frequency specifications.

Alternatively, if the noise and signal frequency bands overlap, but a reference source for the noise is present and the noise is additive, then an adaptive filter may be used to minimize the noise. An adaptive filter possesses a structure that is adjustable in such a way that its performance improves through contact with its environment. Finally, if the noise and signal frequency bands overlap, but the noise is minimal when transformed to another domain, then time–frequency or time–scale analysis may be used to minimize the noise. After transformation, important signal characteristics may be recovered (Baura 2002b).

2.2.1   Frequency Selective Filters

Frequency-selective filtering refers to noise minimization within a certain frequency band. As stated above, this simplest type of filtering is useful when the noise and physiologic signal frequencies do not overlap. Ideally, a frequency-selective filter passes only frequencies of interest, based on a cutoff frequency, ω_c. Each filter form is named for the frequency range that is passed: lowpass filter, highpass filter, bandpass filter, bandstop (all frequencies except a certain band or range) filter (Fig. 2.5).

This linear, time-invariant filter has the form of a subset of the autoregressive moving average, exogenous input (ARMAX) model, which consists of the linear constant coefficient

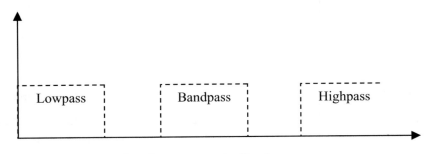

FIGURE 2.5: Schematic of frequency ranges that are passed for various frequency-selective filters

difference equation:

$$\sum_{n=0}^{N} a_n y(k-n) = \sum_{n=0}^{M} b_n u(k-n) + \sum_{n=0}^{P} c_n e(k-n), \qquad (2.2)$$

where $e(k)$ is a sequence of independent and identically distributed random variables with zero mean, otherwise known as white noise, and $a_0 = c_0 = 1$. The model is autoregressive because it looks back upon past values of itself, $y(k-n)$. It possesses a moving average (the $e(k)$ terms), and an exogenous, or external, input, $u(k)$.

For our filter, let us assume that white noise is not present and restrict ourselves to input and output terms:

$$\sum_{n=0}^{N} a_n y(k-n) = \sum_{n=0}^{M} b_n u(k-n). \qquad (2.3)$$

Using unit delays and the z-transform time-shifting property, we can take the z-transform on both sides as:

$$\sum_{n=0}^{N} a_n z^{-n} Y(z) = \sum_{n=0}^{M} b_n z^{-n} U(z). \qquad (2.4)$$

The corresponding transform function, or ratio of output to input in the frequency domain, is:

$$H(z) = \frac{Y(z)}{U(z)} = \frac{\sum_{n=0}^{M} b_n z^{-n}}{\sum_{n=0}^{N} a_n z^{-n}} = \frac{b_0 \left(1 + \frac{b_1}{b_0} z^{-1} + \cdots + \frac{b_M}{b_0} z^{-M}\right)}{1 + a_1 z^{-1} + \cdots + a_N z^{-N}}, \qquad (2.5)$$

recalling that $a_0 = 1$.

If the filter transfer function has coefficients such that $M \leq N$, then it is called an infinite impulse response (IIR) filter. Its name is based on the recognition that long division of the transfer function numerator by its denominator generates an infinite number of terms. If $N = 0$, then the filter is a finite impulse response (FIR) filter because long division of the numerator by denominator obviously contains a finite number of terms. Designers choose FIR systems to specify exactly linear phase or generally linear phase, or choose IIR systems for closed form equations, which are more efficient (lower order) to implement.

Because these filters are rarely designed from scratch, we will not describe the processes for determining the coefficients, a_n and b_n, nor orders, M and N. For design information, the reader is referred to (Oppenheim et al. 1999).

2.2.2   Pseudorandom Binary Sequence

For the frequency-selective filter, we passed a specified frequency range that contains the physiologic signal of interest. If the majority of system noise resides outside this range, then the system signal-to-noise ratio (SNR) is increased. Another filter that increases the SNR is the pseudorandom binary sequence (PRBS) filter. This time, we amplify the signal strength, while leaving the noise constant. Given an input signal that is periodic, the PRBS filter functions as a bandpass filter that does not require frequency specifications.

A PRBS is composed of two values that appear to be randomly introduced but are reproducible by deterministic means. The PRBS filter is one of many spread-spectrum techniques originally developed for military communications but now finding applications in the commercial arena, particularly in cellular telephony. Spread spectrum refers to a modulation technique that spreads the spectrum of a signal by using a very wideband spreading signal. The spreading signal, or encoder, is chosen to have properties that facilitate demodulation, or decoding, of the transmitted signal by the intended receiver. The encoder is also chosen to make demodulation by an unintended receiver as difficult as possible. Since system noise is added in the transmission medium after encoding, the noise can be filtered. The PRBS filtering process is shown in Fig. 2.6.

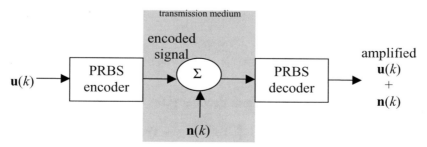

FIGURE 2.6: PRBS filtering process

TABLE 2.1: *M*-sequences		
N	*M*	*M*-SEQUENCE
2	3	110
3	7	1110100
4	15	111101011001000
5	31	1111100110100100001010111011000

2.2.2.1 PRBS Properties

The PRB sequence consists of a predetermined binary sequence of ones and zeros of maximal length, $M = 2^N - 1$, where N is the filter order. These maximal length sequences, or M-sequences, possess several useful properties:

1. An M-sequence contains $1/2(N + 1)$ ones and $[1/2(N + 1) - 1]$ zeros.
2. The modulo-2, or binary, sum of an M-sequence and any phase shift of the same sequence is another phase of the same M-sequence.
3. The periodic autocorrelation function, $\phi_{\text{mm}}(k)$, has two values and is given by

$$\phi_{\text{mm}}(k) = \begin{cases} 1.0, & k = iM \\ \frac{-1}{M}, & \text{otherwise} \end{cases}, \tag{2.6}$$

where i is any integer and M is the sequence period. The M-sequences for $N \leq 5$ are listed in Table 2.1.

2.2.2.2 Filter Construction

As shown in Fig. 2.6, the PRBS filter is composed of an encoder and decoder. The encoder, $e(k)$, is merely the original M-sequence. The decoder, $d(k)$, is a similar code in which each 0 has been replaced by -1. We can take the cross-correlation of the encoder and decoder, which is calculated as:

$$\phi_{\text{ed}}(k) = \sum_{n=-\infty}^{\infty} e(n)\, d(k + n). \tag{2.7}$$

Cross-correlation of our encoder and decoder gives the unusual result of:

$$\phi_{\text{ed}}(k) = \begin{cases} \frac{M+1}{2}, & k = 0 \\ 0, & \text{otherwise} \end{cases}. \tag{2.8}$$

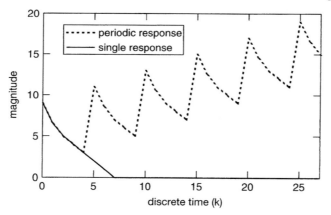

FIGURE 2.7: Stackup phenomenon

This cross-correlation can be used to filter a periodic signal. Each incoming period, τ, of the input signal is encoded, that is, multiplied by the next value of the encoder sequence. The encoded signal is then passed through its transmission medium, which may add system noise. The received signal is decoded by periodic convolution with the decoder sequence, such that

$$decoder\_output(k) = \sum_{i=0}^{M-1} d(n) \cdot encoder\_output(i\tau + k). \qquad (2.9)$$

As per Eq. (2.8), the decoder output will be composed of an amplified periodic signal and the original level of transmitted noise.

While not obvious, this filter also reduces the signal stackup. Stackup occurs when the time decay of a perturbed signal response back to its original DC value is longer than its period (Fig. 2.7). In other words, the onset of a periodic signal, which results from a periodic perturbation, occurs faster than its time delay. The PRBS filter eliminates stackup because only one perturbation response is preserved.

2.2.2.3 PRBS Example

As an example, let us calculate the filter response of a periodic sequence $\{5\ 3\ 2\ 0\}$ ($\tau = 4$, Fig. 2.8) to a second-order filter, with $M = 3$.

From Table 2.1, we know the encoder sequence, $e(k)$, is $\{1\ 1\ 0\}$. Multiplying each period by the encoder sequence results in the encoder output of $\{5\ 3\ 2\ 0\ 5\ 3\ 2\ 0\ 0\ 0\ 0\ 0\ 5\ 3\ 2\ 0\ 5\ 3\ 2\ 0\ 0\ 0\ 0\ 0\ 0\}$. Convolving this encoder output with the decoder sequence ($\{1\ 1\ -1\}$) results in the decoder output $\{10\ 6\ 4\ 0\ 0\ 0\ 0\ 0\ 0\ 0\ 0\ 0\}$. Note that the decoder output is merely an amplification of the input sequence.

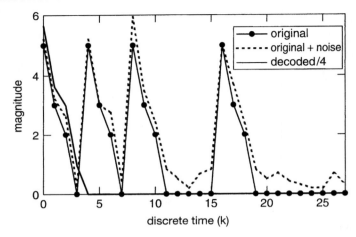

FIGURE 2.8: PRBS filter results for $M = 7$

Now let us add uniform noise in the range $\{0\ 1\}$ to the encoder output of this sequence in a third-order filter, with $M = 7$ (Fig. 2.8). This noise significantly degrades the encoder output sequence. However, if we decode the output (Fig. 2.8), the final result is now closer to the original sequence for the first period. Although the decoded sequence, if normalized, still does not equal the original, the noise offset is now more uniform.

In a similar manner, IVAC Corporation (now part of Cardinal Health) used a PRB sequence of length $M = 31$ in the Signature Edition large volume infusion pump. This pump calculates catheter resistance, as a first step to detection of infiltration. Infiltration is a drug-infusion complication during which an intravenous solution or medication is infused into the tissue surrounding a blood vessel. Theoretically, resistance is calculated from the ratio of catheter pressure to infusion pump flow. However, in practice, the pressure waveform may possess significant noise artifact due to the patient's own blood pressure or other external sources. External sources include eating with a catheterized arm and wheelchair self-ambulation.

The infusion pump flow packets were encoded using the PRB sequence. The measured pressure packets were then decoded using PRBS (Fig. 2.9).

The higher signal-to-noise ratio in the pressure waveform enabled accurate calculations of resistance, as well as a decrease in false alarms based on pressure settings (Voss et al. 1997).

2.2.3   Adaptive Filters

Often, a signal and noise exist in the same frequency range, rendering frequency-selective filtering ineffective. However, under certain constraints, an adaptive filter may be utilized. An adaptive filter possesses a structure that is adjustable in such a way that its performance improves through contact with its environment. Such behavior is the result of several constraints on the

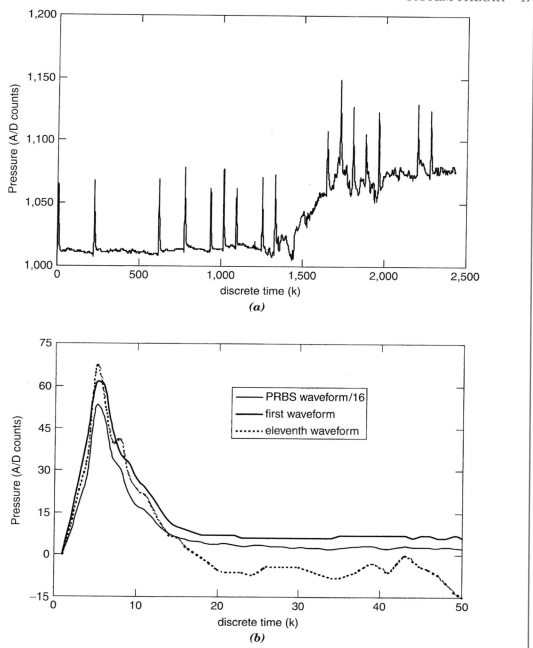

FIGURE 2.9: (a) Typical raw pressure during wheelchair self-ambulation, at a flow rate of 50 ml/h. (b) The first raw waveform, the eleventh waveform, and the PRBS filter output scaled by 1/16 are shown. For comparison purposes, the baseline value of each waveform has been subtracted. Courtesy of Cardinal Health, San Diego, CA

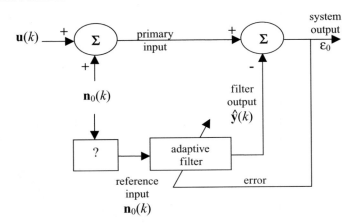

FIGURE 2.10: Adaptive noise canceller

signal, $u(k)$; the noise associated with the signal, $n_0(k)$; the reference noise source, $n_1(k)$; and the filter output, $\hat{y}(k)$. When an adaptive filter is used as an adaptive noise canceller, the reference noise source is filtered and subtracted from the primary input containing the signal and noise to eliminate the noise by cancellation. The adaptive filtering process, used for noise cancellation, is shown in Fig. 2.10.

2.2.3.1 Adaptive Filter Properties

The system constraints for an adaptive filter are as follows:

1. $u(k)$, $n_0(k)$, $n_1(k)$, and $\hat{y}(k)$ are statistically stationary (not variable with time).
2. $u(k)$, $n_0(k)$, $n_1(k)$, and $\hat{y}(k)$ have zero means.
3. $u(k)$ is uncorrelated with $n_0(k)$ and $n_1(k)$.
4. $n_0(k)$ is correlated with $n_1(k)$.

By feeding the system output back to an adaptive filter and adjusting the filter to minimize system output power (i.e., the power of the system error), the output becomes a best approximation to the signal.

2.2.3.2 Filter Construction

For our adaptive noise canceller, let us use an FIR construction and Widrow and Hoff's least mean squares (LMS) algorithm for identification of the $b_n(k)$ coefficients (Widrow and Hoff 1960, Widrow and Sterns 1985):

$$\hat{y}(k) = \sum_{n=0}^{M} b_n(k)u(k-n). \tag{2.10}$$

Note that our filter coefficients vary with time, unlike coefficients in the standard FIR filter. Let us rewrite our filter so that the inputs are coefficients are vectors of length M:

$$\hat{y}(k) = \mathbf{b}^T(k)\mathbf{u}(k).\qquad(2.11)$$

As the name of the algorithm implies, we will use a performance function, $\xi(k)$, that is the expected value of the mean instantaneous squared error, $\varepsilon_0^2(k)$, between the desired and filtered output:

$$\xi[\mathbf{b}(k)] = E\{\varepsilon_0^2(k)\} = E\{y(k) - \hat{y}(k)\}^2.\qquad(2.12)$$

The performance function enables us to devise a method for updating the coefficients, which Widrow calls weights. First, we substitute Eq. (2.11) into Eq. (2.12), noting that the expected value of any sum is the sum of expected values:

$$\xi[\mathbf{b}(k)] = E\{y^2(k)\} + E\{\mathbf{b}^T(k)\mathbf{u}(k)\mathbf{u}^T(k)\mathbf{b}(k)\} - E\{2y(k)\mathbf{u}^T(k)\mathbf{b}(k)\}.\qquad(2.13)$$

Next, we assume that these are statistically stationary, and that the weights are no longer adjusted (independent of time, having reached steady state). Because the expected value of a product is the product of expected values when variables are statistically stationary, the performance function can be simplified to:

$$\xi(\mathbf{b}) = E\{y^2(k)\} + \mathbf{b}^T E\{\mathbf{u}(k)\mathbf{u}^T(k)\}\mathbf{b} - 2E\{y(k)\mathbf{u}^T(k)\}\mathbf{b}.\qquad(2.14)$$

We further simplify Eq. (2.14) by defining $\underline{\mathbf{R}}(k)$ as the square matrix:

$$\underline{\mathbf{R}}(k) = E\{\mathbf{u}(k)\mathbf{u}^T(k)\}\qquad(2.15)$$

$$= E\left\{\begin{matrix} u^2(k) & u(k)u(k-1) & \cdots & u(k)u(k-M) \\ u(k-1)u(k) & u^2(k-1) & \cdots & u(k-1)u(k-M) \\ \vdots & \vdots & \ddots & \vdots \\ u(k-M)u(k) & u(k-M)u(k-1) & \cdots & u^2(k-M) \end{matrix}\right\}.\qquad(2.16)$$

We also define $\mathbf{p}(k)$ as the column vector:

$$\mathbf{p}(k) = E\{y(k)\mathbf{u}(k)\} = E\{y(k)u(k) \quad y(k)u(k-1)\dots y(k)u(k-M)\}^T.\qquad(2.17)$$

When $\mathbf{u}(k)$ and $y(k)$ are stationary, the elements of both $\underline{\mathbf{R}}(k)$ and $\mathbf{p}(k)$ are all constant second-order statistics. Substituting Eqs. (2.15) and (2.17) into (2.14), we obtain the simplified performance function

$$\xi(\mathbf{b}) = E\{y^2(k)\} + \mathbf{b}^T\underline{\mathbf{R}}(k)\mathbf{b} - 2\mathbf{p}(k)^T\mathbf{b}.\qquad(2.18)$$

Note that this performance function is quadratic, with a bowl-shaped performance surface that is concave upward. Therefore, only a single global minimum exists, and no local minima are present.

To find this global minimum, we use the method of steepest descent for optimization, which requires gradient estimation. The gradient is a vector consisting of partial derivatives of the performance function. Widrow and Hoff assumed that $\varepsilon_0^2(k)$ is an estimate of $E\{\varepsilon_0^2(k)\}$. Returning to Eqs. (2.11) and (2.12), we calculate the estimated gradient as

$$\hat{\nabla}\xi[\mathbf{b}(k)] = \frac{\partial\hat{\xi}[\mathbf{b}(k)]}{\partial\mathbf{b}(k)} = \frac{\partial}{\partial\mathbf{b}(k)}\varepsilon_0^2(k) \qquad (2.19)$$

$$\hat{\nabla}\xi[\mathbf{b}(k)] = \frac{\partial}{\partial\mathbf{b}(k)}[y(k) - \mathbf{b}^T(k)\mathbf{u}(k)]^2 \qquad (2.20)$$

$$\hat{\nabla}\xi[\mathbf{b}(k)] = -2\varepsilon_0^2(k)\mathbf{u}(k) \qquad (2.21)$$

We can now specify the LMS algorithm as iterative updating of the weights by steepest changes in the performance function, which is Eq. (2.21):

$$\mathbf{b}(k+1) = \mathbf{b}(k) - 2\mu\varepsilon_0^2(k)\mathbf{u}(k). \qquad (2.22)$$

Note that μ is a gain constant that regulates the speed and stability of adaptation. The LMS algorithm is guaranteed to converge to the optimal solution only if the inverse of the maximum eigenvalue of $\underline{\mathbf{R}}(k)$ is greater than the gain constant, which in turn must be greater than zero:

$$\frac{1}{\lambda_{max}} > \mu > 0. \qquad (2.23)$$

Convergence of the weights is slow. Additionally, the algorithm is sensitive to the eigenvalue spread, which is the ratio of the largest to smallest eigenvalue.

2.2.3.3 Adaptive Filter Example

The classic biomedical example of adaptive filtering is Widrow's cancellation of the maternal heartbeat in recorded fetal electrocardiography (ECG) (Widrow et al. 1975). Although this example uses more than one reference noise input, which is beyond the scope of our general discussion, the results are readily understood.

During recording of fetal ECG, interference results from the maternal heartbeat, which has an amplitude two to ten times greater than that of the fetal heartbeat. Interference also results from the background noise of muscle activity and fetal motion, with an amplitude greater than or equal to that of the fetal heartbeat. To minimize this interference, Widrow's group used four chest leads to record the maternal heartbeat and other multiple reference inputs for the canceller. A single abdominal lead recorded the combined maternal and fetal heartbeats

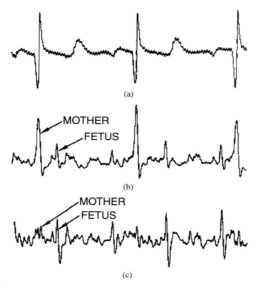

FIGURE 2.11: Result of adaptive noise cancellation within fetal ECG. (a) Reference noise input (chest lead). (b) Primary input (abdominal lead). (c) Noise canceller output. Reprinted from (Widrow, 1975), © 2007 IEEE.

that served as a primary input. The data were prefiltered with a bandwidth of 0.3–75 Hz, digitized with a sampling rate of 512 Hz, and recorded on tape. The recorded data were input to a multichannel LMS adaptive filter. Each of the reference inputs was processed with 32 coefficients with nonuniform (log periodic) spacing and a total delay of 129 ms.

As shown in Fig. 2.11(b), baseline drift and 60 Hz interference are clearly observable in the primary input obtained from the abdominal lead. Further, the maternal heartbeat (Fig. 2.11(a)) dominates in the primary input recording. The maternal heartbeat and three other chest recordings containing the 60 Hz interference served as a reference to reduce these interferences. In the filter output (Fig. 2.11(c)), the fetal heartbeat is clearly discernable (Widrow et al. 1975).

2.2.4 Wavelet Transforms

If a signal and noise exist in the same frequency range, but a reference noise source in a stationary system is not available for adaptive filtering, the signal of interest may still be recoverable. In this case, data may be transformed to recover important signal characteristics using a time–scale or time–frequency distribution. In this brief review, we only cover time–scale distributions, as they are easier to compute and implement in medical instruments. Time–scale distributions are better known as wavelet transforms.

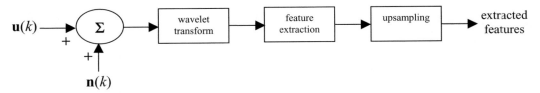

FIGURE 2.12: Wavelet transform denoising

A time–scale distribution is a two-dimensional mapping of the fraction of the energy of a one-dimensional signal at time, t, and scale, a. Scale is a physical attribute representing compression. Time–scale distributions possess an inherent denoising feature that minimizes noise artifact. By transforming a signal and noise to an alternate scale, important signal features may be extracted. The recovered features must then be upsampled to the original scale, for proper positioning of the features within the original signal. This denoising process is illustrated in Fig. 2.12.

2.2.4.1 Wavelet Transform Properties

By definition, a wavelet, $\varphi(t)$, is a function with a mean of zero, such that

$$\int \varphi(t)\, dt = 0. \tag{2.24}$$

Each wavelet function may be used to bandpass filter a signal. Using this method, the scale plays the role of a local frequency. As a increases, wavelets are stretched and interact with low frequencies. As a decreases, wavelets are compressed and interact with high frequencies. The continuous wavelet transform, $WT(t, a)$, of a signal, $u(t)$, can be defined in the time domain as

$$WT(t, a) = \frac{1}{\sqrt{a}} \int u(\tau)\varphi * \left(\frac{\tau - t}{a} \right) d\tau. \tag{2.25}$$

The center frequency and bandwidth of the transform vary inversely with scale, such that the ratio of the center frequency to the bandwidth is constant.

For efficient processing, only dyadic scales are used. Dyadic refers to powers of the numeral 2. Using direct substitution, this leads us to the dyadic discrete wavelet transform, $WT(k, 2^j)$:

$$WT(k, 2^j) = \frac{1}{\sqrt{2^j}} \sum_{i=0}^{N-1} u(i)\varphi * \left(\frac{i - k}{2^j} \right). \tag{2.26}$$

The calculations in Eq. (2.26) may be classified functionally as filtering through convolution and downsampling (retaining every other sample).

2.2.4.2 Filter Construction

By choosing a wavelet or scaling filter for $\varphi(k)$ in Eq. (2.26), lowpass or highpass filtering may be accomplished. Two common wavelet and scaling filter pairs are given in Table 2.2.

When a scaling filter, $\varphi_a(k)$, is used, convolution results in smoothing or lowpass filtering. The resulting wavelet transform is composed of approximation coefficients. As shown in

TABLE 2.2: Haar and Daubechies D4 wavelet and scaling filters

Haar wavelet filter $\varphi_d(k) = \{ -0.7071, +0.7071\}$	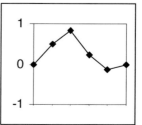
Haar wavelet filter $\varphi_d(k) = \{ -0.7071, +0.7071\}$	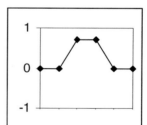
Daubechies D4 wavelet filter $\varphi_d(k) = \{-0.1294, -0.2241, 0.8365, -0.4830\}$	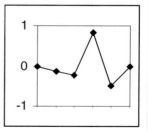
Daubechies D4 scaling filter $\varphi_a(k) = \{0.4830, 0.8365, 0.2241, -0.1294\}$	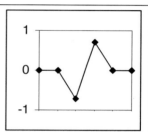

Fig. 2.13, an ECG beat (Sugimachi 1992) degrades in subsequent scales of Haar approximation coefficients to emphasize the slower T wave, rather than the faster QRS complex.

When a wavelet filter, $\varphi_d(k)$, is used, convolution results in emphasis of discontinuities or highpass filtering. The resulting wavelet transform is composed of detail coefficients. As shown in Fig. 2.13, the same electrocardiogram beat degrades in subsequent scales of Haar detail coefficients to emphasize the high-frequency QRS complex.

The choice of an optimum wavelet or scaling filter for a particular application is an art, not a science. This process involves much trial and error, using an appropriate sample of

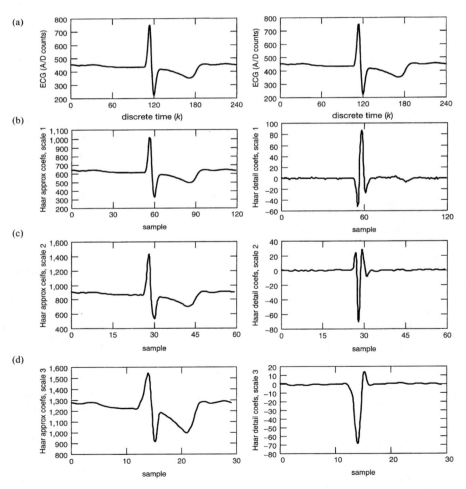

FIGURE 2.13: Electrocardiogram beat (Sugamachi 1992), with wavelet transforms. (a) Original beat. (b) Scale 1 Haar approximation and detail coefficients. (c) Scale 2 Haar approximation and detail coefficients. (d) Scale 3 Haar approximation and detail coefficients

data to search for consistent feature extraction. In addition to the wavelet or scaling filter, the appropriate scale must also be chosen.

2.2.4.3 Wavelet Transform Example

In the 1990s, Boudreaux-Bartels's group investigated a wavelet-based QRS complex detector. The detection was based on a cubic spline mother wavelet they designed, with a center frequency equal to 120 Hz and a bandwidth of 240 Hz. Inspecting ECG detail coefficients at scales 2 and 3, they identified the R wave of a QRS complex as at least 60% of the maximum value of the detail coefficients, if found at both scales. QRS detection is illustrated in Fig. 2.14.

Algorithm validation was accomplished using ECG beats from the American Heart Association (AHA) database. The database includes reference beat notations from cardiologists. 1950 beats from channel 1 of the 30 min Tape 3203 were analyzed, using this method and

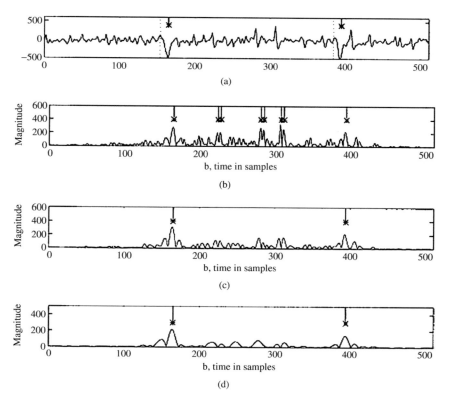

FIGURE 2.14: Electrocardiograms with external noise from American Heart Association Tape 3203 and corresponding detail coefficients. (a) ECG, with dashed lines indicating cardiologist estimate of QRS onset and tic marks indicating wavelet estimate of R wave locations. (b) Detail coefficients at scale 1. (c) Detail coefficients at scale 2. (d) Detail coefficients at scale 3 (Kadambe, 1999). © 2007 IEEE.

three traditional time-based methods. Error rate was defined as the sum of false positives and false negatives, divided by the total number of beats.

Using the cubic spline wavelet, an error rate of 3% was obtained. In contrast, the Hamilton–Tompkins, Multiplication of Backward Difference, and Okada algorithms resulted in error rates of 6, 15, and 34%, respectively (Murray 1994). In a subsequent analysis of four AHA tapes (including tape 3203), representing 8598 beats with various arrhythmias, the wavelet method results in a mean error rate of 7%, while the time-based methods resulted in mean error rates of 12, 15, and 21%, respectively (Kadambe 1999).

2.3   MODELING

Once a low noise approximation of the true physiologic signal is obtained, system identification, or modeling, can be performed to estimate the system operator. Typical reasons for modeling the system operator include classification, prediction of future behavior, or insight into underlying physiologic mechanisms.

Modeling techniques differ for signals that are evenly sampled in time, versus those with sparse data sampling. If an evenly sampled system is linear, that is, defined by a linear differential equation, it may be modeled using the classic autoregressive moving average exogenous input (ARMAX) model or one of its variations. If the system constraints are relaxed to that of a multiple input, multiple output system that is nonlinear but still time-invariant, then the system may be described as an artificial neural network. If the system constraints are further relaxed so that the system operator is nonlinear, time-invariant, and sufficiently complex, it may not be easily described by mathematical equations. Given such a system, the system operator may be described using fuzzy logic (Baura 2002b).

For a sparsely sampled signal, linear or nonlinear compartmental modeling is used (Baura 2002b). Compartmental modeling is not included in this Lecture because it is not typically incorporated into market-released patient monitoring and diagnostic devices. However, fuzzy control, which is closely related to fuzzy models, is included in this chapter. Because of the inherent nonlinearity of many physiologic signals, they are more easily controlled by fuzzy control, than classic control.

2.3.1   The ARMAX Model

Linear modeling is the simplest approach to classification or prediction of future behavior. We already introduced the ARMAX model in Eq. (2.2):

$$\sum_{n=0}^{N} a_n y(k-n) = \sum_{n=0}^{M} b_n u(k-n) + \sum_{n=0}^{P} c_n e(k-n), \qquad (2.2)$$

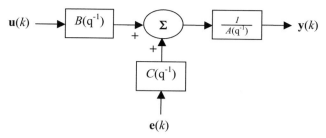

FIGURE 2.15: ARMAX system

where $e(k)$ is a sequence of independent and identically distributed random variables with zero mean, otherwise known as white noise, and $a_0 = c_0 = 1$. The model is autoregressive (AR) because it looks back upon past values of itself, $y(k - n)$. It possesses a moving average (MA, the $e(k)$ terms), and an exogenous (X), or external, input, $u(k)$.

Alternatively, Eq. (2.2) may be represented using the argument q^{-1}, which denotes the backward shift operator. Using this argument, a delay of one sample in the input, $u(k)$, may be represented as

$$u(k - 1) = q^{-1}u(k). \tag{2.27}$$

Using the backward shift operator, we may rewrite Eq. (2.2) as

$$A(q^{-1})y(k) = B(q^{-1})u(k) + C(q^{-1})e(k), \tag{2.28}$$

where

$$A(q^{-1}) = 1 + a_1q^{-1} + \cdots + a_Nq^{-N}, \tag{2.29}$$

$$B(q^{-1}) = b_0 + b_1q^{-1} + \cdots + b_Mq^{-M}, \tag{2.30}$$

$$C(q^{-1}) = 1 + c_1q^{-1} + \cdots + c_Pq^{-P}. \tag{2.31}$$

The parameter vector, θ, for this model consists of

$$\theta = [a_1 \ldots a_N \quad b_0 \ldots b_M \quad c_0 \ldots c_P]^T. \tag{2.32}$$

This system is illustrated in Fig. 2.15.

An important variation of the ARMAX model is the ARX, or controlled autoregressive, model:

$$A(q^{-1})y(k) = B(q^{-1})u(k) + e(k). \tag{2.33}$$

The corresponding parameter vector for this model is

$$\boldsymbol{\theta} = [a_1 \ldots a_N \quad b_0 \ldots b_M]^T. \tag{2.34}$$

We can rewrite Eq. (2.33) as

$$y(k) = \boldsymbol{\phi}^T(k)\boldsymbol{\theta} + e(k), \tag{2.35}$$

where the regression vector, $\boldsymbol{\phi}^T(k)$, is

$$\boldsymbol{\phi}^T(k) = [-y(k-1)\ldots -y(k-N)u(k)\ldots u(k-M)]^T. \tag{2.36}$$

2.3.1.1 Model Identifiability

Model identifiability refers to the possibility of theoretically and practically obtaining unique estimates of all the unknown model parameters. The ARMAX and ARX models are both theoretically identifiable. However, the practical process of parameter estimation may not yield acceptable results. Problems may arise from different types of experimental error, the number of data points, and the true system, leading to a loss of practical identifiability.

A model may be uniquely identifiable, nonuniquely identifiable, or nonidentifiable. If it is uniquely identifiable, the parameters may be uniquely determined. If it is nonuniquely identifiable, one or more of the parameters possesses more than one, but a finite, number of possible values. If it is nonidentifiable, one or more parameters possess an infinite number of solutions.

2.3.1.2 Parameter Estimation

Given an identifiable model, the model parameters may be calculated through a performance function, $\xi(\boldsymbol{\theta})$. Let us use the prediction error to derive a prediction error method for estimating the parameter vector.

The error, $\varepsilon(k)$, between an output, $y(k)$, and its estimate, $\hat{y}(k)$, is

$$\varepsilon(k) = y(k) - \hat{y}(k) \tag{2.37}$$

Since the estimate may be modeled as

$$\hat{y}(k) = \boldsymbol{\phi}^T(k)\boldsymbol{\theta}, \tag{2.38}$$

the error may be calculated as

$$\varepsilon(k) = y(k) - \boldsymbol{\phi}^T(k)\boldsymbol{\theta}. \tag{2.39}$$

Over time, the error vector containing these residuals is defined as

$$\varepsilon(\mathbf{k}) = [\varepsilon(1)\ldots \varepsilon(k)], \tag{2.40}$$

where K is $> (N+M+1)$, the dimension of the parameter vector. Since there are $(N+M+1)$ parameters in $\boldsymbol{\theta}$, it should be theoretically possible to solve for $\boldsymbol{\theta}$ from $(N+M+1)$ measurements. However, $K > (N+M+1)$ measurements are used to account for noise, disturbances, and model misfit.

The performance functions utilized in prediction error methods possess the form

$$\xi(\boldsymbol{\theta}) = h[V(\boldsymbol{\theta})], \tag{2.41}$$

where $h(k)$ is a scalar-valued function that must satisfy certain conditions. In particular, the least squares estimate of $\boldsymbol{\theta}$ is defined as the vector that minimizes the mean squared error performance function

$$\xi(\boldsymbol{\theta}) = \frac{1}{2} \cdot \frac{1}{K} \sum_{k=1}^{K} \varepsilon^2(k). \tag{2.42}$$

The factor of $1/2$ has been added to simplify calculation of the performance function derivative.

The minimum of the mean squared error performance function is calculated by setting its derivative equal to zero:

$$\frac{d\xi(\boldsymbol{\theta})}{d\boldsymbol{\theta}} = \frac{d}{d\boldsymbol{\theta}} \left\{ \frac{1}{2} \cdot \frac{1}{K} \sum_{k=1}^{K} \varepsilon^2(k) \right\} = 0. \tag{2.43}$$

Substituting Eq. (2.39) into Eq. (2.43) yields

$$\frac{d\xi(\boldsymbol{\theta})}{d\boldsymbol{\theta}} = \frac{d}{d\boldsymbol{\theta}} \left\{ \frac{1}{2} \cdot \frac{1}{K} \sum_{k=1}^{K} [y(k) - \boldsymbol{\phi}^T(k)\boldsymbol{\theta}]^2 \right\} = 0 \tag{2.44}$$

$$\frac{1}{K} \sum_{k=1}^{K} [\boldsymbol{\phi}(k)y(k) - \boldsymbol{\phi}(k)\boldsymbol{\phi}^T(k)\boldsymbol{\theta}] = 0. \tag{2.45}$$

Moving the negative terms to the other side of the equation results in

$$\frac{1}{K} \sum_{k=1}^{K} \boldsymbol{\phi}(k)y(k) = \frac{1}{K} \sum_{k=1}^{K} \boldsymbol{\phi}(k)\boldsymbol{\phi}^T(k)\,\boldsymbol{\theta}. \tag{2.46}$$

We can then solve for $\hat{\boldsymbol{\theta}}$, the least squares estimate, as

$$\hat{\boldsymbol{\theta}} = \left[\frac{1}{K} \sum_{k=1}^{K} \boldsymbol{\phi}(k)\boldsymbol{\phi}^T(k) \right]^{-1} \frac{1}{K} \sum_{k=1}^{K} \boldsymbol{\phi}(k)y(k). \tag{2.47}$$

2.3.1.3 Model Validation

Given various models for the same data sets, how do we choose the best model? Model validation involves analyzing the results of parameter identification to select an "optimum" model. For linear models, we evaluate the coefficient of variation (CV), goodness of fit, and residual statistics.

The coefficients of variation are calculated from the covariance matrix, $\mathbf{V}[K]$, which is the expected value of the regression vector, multiplied by its transpose:

$$\mathbf{V}[K] = \frac{1}{K} \sum_{k=1}^{K} \boldsymbol{\phi}(k)\boldsymbol{\phi}^{\mathrm{T}}(k). \qquad (2.48)$$

The diagonal elements of the covariance matrix contain estimates of the variance associated with each identified parameter. The square roots of these variances are used to calculate the standard deviations and therefore associated CV for each parameter estimates. CV is also known as the fractional standard deviation (FSD), and is merely the ratio of a standard deviation to its mean value. When the CVs of estimated parameter values are unreasonably large (i.e., much greater than 100%), the model may be considered suboptimal. Large CVs may arise from limitations in the experimental data, such as a small number of measurements or large measurement errors. Large CVs may also arise from utilization of a model that is too complex for the available experimental data. As the CVs become larger, the covariance matrix tends toward nonsingularity (nonunique identifiability).

Goodness of fit may be determined using Akaike's final prediction error criterion (FPE) (Akaike 1970). Because a more complicated model with more parameters may better fit experimental data, the number of model parameters is weighed against the number of data points and mean squared error, which refers to the average of the squared difference between the observed and estimated data:

$$\text{FPE} = \left[\frac{NN+p}{NN-p}\right]\left\{\frac{1}{NN} \sum_{k=1}^{NN} [y(k) - \hat{y}(k)]^2\right\}, \qquad (2.49)$$

where NN is the number of data points and p is the number of parameters. In terms of this criterion, the best model is that which yields the lowest value of FPE.

Residual statistics refers to estimation of the system noise. If this noise is assumed to be white, Gaussian, and of zero mean, then the residuals should display these properties. If the residuals do not meet the assumptions, a systematic error in model identification may be present.

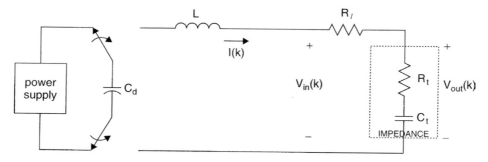

FIGURE 2.16: Damped sinusoid circuit for external defibrillation

2.3.1.4 ARMAX Model Example

In the 1990s, many researchers began to investigate alternate waveform shapes for automated external defibrillation. While the traditional damped sinusoid waveform was known to be effective, it required large current magnitudes during discharge. For a nominal 50 Ω patient, 60 A was transmitted at the highest defibrillator setting of 360 J. This high current required a large capacitor and large inductor for waveform shaping (Fig. 2.16). Indeed, a typical external defibrillator during this timeframe weighed approximately 20 lbs, which limits its utility for automated applications outside the hospital environment.

Various research groups reported defibrillation efficacy for different waveform shapes. Defibrillation efficacy refers to the percentage of successful defibrillation attempts at a particular voltage. Generally, V_{50}, the total voltage required for 50% successful defibrillation, was reported in published studies. Baura hypothesized that an accurate estimate of transthoracic impedance could be used to predict V_{50} for various waveform shapes. While a nominal 50 Ω resistance was typically used to describe transthoracic impedance, researchers had known since at least the 1970s that transthoracic impedance was not purely resistive. For example, simultaneous plots of canine defibrillation voltage versus current result in an ellipse, rather than straight line (Fig. 2.17).

Transthoracic impedance was modeled as a resistor, R_t, and capacitor, C_t, in series. This impedance was connected to a defibrillator damped sinusoid waveform (HP43110), $V_{in}(k)$, and load resistor, R_l (Fig. 2.16). The transfer function between $V_{in}(z)$ and $V_{out}(z)$, the voltage across R_t and C_t, was derived from circuit analysis as:

$$\frac{V_{out}(z)}{V_{in}(z)} = \frac{\frac{R_t + \frac{1}{C_t}}{R_t + R_l + \frac{1}{C_t}} + \frac{\frac{1}{C_t} - R_t}{R_t + R_l + \frac{1}{C_t}} z^{-1}}{1 + \frac{\frac{1}{C_t} - R_t - R_l}{R_t + R_l + \frac{1}{C_t}} z^{-1}}. \qquad (2.50)$$

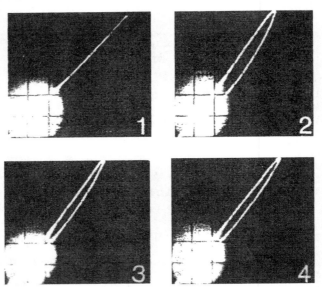

FIGURE 2.17: Simultaneous $X-Y$ plot of current (10A per division) and voltage (500 V per division) from defibrillator discharge. Dog G: (1) Discharge into artificial resistive load of 50 Ω; (2) first transthoracic discharge; (3) second transthoracic discharge; (4) tenth transthoracic discharge. Reprinted from (Dahl, 1976) with permission of C. F. Dahl.

To derive an equivalent transfer function, an ARX model was used to model the impulse response between the measured voltage source and impedance as

$$V_{\text{outm}}(k) = -a_1 V_{\text{outm}}(k-1) + b_0 V_{\text{inm}}(k) + b_1 V_{\text{inm}}(k-1). \qquad (2.51)$$

The model coefficients were obtained from measured $V_{\text{inm}}(k)$ and $V_{\text{outm}}(k)$ using least squares estimation. Substitution of Eq. (2.51) into Eq. (2.47) yields

$$\hat{\boldsymbol{\theta}} = \begin{bmatrix} a_1 & b_0 & b_1 \end{bmatrix}^T = \left[\frac{1}{K} \sum_{k=0}^{K-1} \boldsymbol{\phi}(k) \boldsymbol{\phi}^T(k) \right]^{-1} \frac{1}{K} \sum_{k=0}^{K-1} \boldsymbol{\phi}(k) V_{\text{outm}}(k), \qquad (2.52)$$

where $K =$ number of samples per waveform period and

$$\boldsymbol{\phi}^T(k) = [-V_{\text{outm}}(k-1) V_{\text{inm}}(k) V_{\text{inm}}(k-1)]^T. \qquad (2.53)$$

1000 samples of each voltage were digitized to 12 bits with sampling frequency 20 kHz (National Instruments LabPC + board), lowpass filtered with a corner frequency of 40 Hz, and downsampled by 8 samples before applying least squared estimation. Transformation of

Eq. (2.50) to the z domain and rearrangement yielded the transfer function

$$\frac{V_{\text{outm}}(z)}{V_{\text{inm}}(z)} = \frac{V_{\text{out}}(z)}{V_{\text{in}}(z)} = \frac{b_0 + b_1 z^{-1}}{1 + a_1 z^{-1}}. \tag{2.54}$$

Equating both transfer functions (2.50) and (2.54) yielded an estimate of the transthoracic resistance

$$R_t = \frac{(b_0 - b_1) R_l}{2(1 - b_0)}. \tag{2.55}$$

For each measurement, the resistance value from a single defibrillation pulse was estimated. Capacitance was estimated by fitting the measured decay of $V_{\text{outm}}(k)$ from its peak value (Fig. 2.18) to two exponentials, using Powell's successive quadratic estimation.

This separate step was necessary since the described ARX method does not have sufficient resolution for accurate capacitance estimates on the order of 10^{-5} Farads.

In five female Yorkshire swine, defibrillation input and output voltages for defibrillation shocks of 200 J were acquired. Mean transthoracic resistance and capacitance were estimated as 35 Ω and 299.4 μF, respectively. A voltage versus current plot, based on estimated values in a constructed RC circuit, was then compared to actual defibrillation voltage versus current plots for each swine. A typical swine transthoracic voltage versus current plot for swine 1003 is shown in Fig. 2.19.

These voltage versus current plot comparisons confirmed the accuracy of the estimation technique (Baura 2000a, 2000b, 2001, 2002b).

2.3.2   Artificial Neural Networks

Physiologic data are often nonlinear. If we allow our system operator, $\mathbf{H}(k)$, in Fig. 2.2 to be nonlinear but time-invariant, we may describe the system operator as an artificial neural network (ANN).

As its name implies, an artificial neural network refers to a mathematical model of human brain processing. Indeed, in the 1940s, physiologists and electrical engineers worked together toward this goal. However, over time, it was discovered that these models did not simulate human neuron processing. The models remain because of their useful properties, such as their ability to "learn" a nonlinear function represented by input–output pairs.

One of the simplest neural network architectures is the multilayer feedforward network. Feedforward networks easily model nonlinear systems, possess a high degree of generalization, and are suitable for parallel processing. However, these networks are slow to train, as the parameter estimates converge slowly. They also cannot be easily analyzed for network behavior.

A multilayer feedforward network contains one layer of inputs, $\mathbf{u}(k)$, one or more hidden layers of neurons, $\mathbf{x}(k)$, and one output layer of neurons, $\hat{\mathbf{y}}(k)$. Ever since Cybenko (1989) and

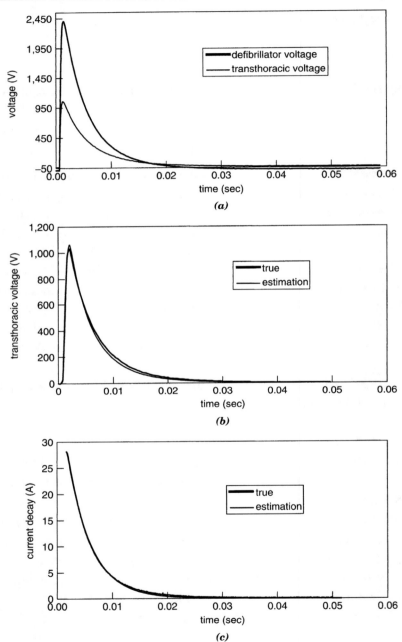

FIGURE 2.18: Estimation of transthoracic resitance and capacitance from typical voltage data (swine 1003). (a) Defibrillator and transthoracic voltage. (b) Estimated and true decimated, lowpass filtered, transthoracic voltage. Resistance = 35 Ω. (c) Estimated and true transthoracic current decay. Capacitance = 355.2 μF. Based on (Baura 2000b)

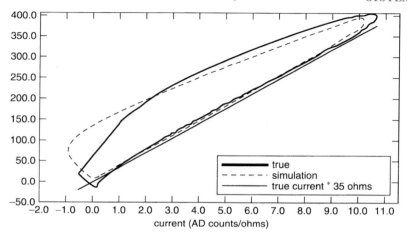

FIGURE 2.19: Typical swine transthoracic voltage versus current plot (swine 1003). The simulation was obtained by passing the scaled, defibrillator voltage from a voltage divider network through a 50 Ω load resistor, 35 Ω resistor, and 357 μF capacitor. The estimated transthoracic resistance and capacitance are 35 Ω and 355.2 μF, respectively. The conversion factor for the observed data was 2.65 volts/AD counts. Without a voltage divider network, the conversion factor for the simulated data was 0.00122 volts/AD counts. The voltage for a transthoracic impedance of 35 Ω alone is also plotted (straight line). Based on (Baura 2000b)

Hornik et al. (1989) proved that one hidden layer is enough to approximate any continuous function, multilayer feedforward networks have typically utilized one hidden layer only. Each weight between the input and hidden layers is represented by $w_{lm}(k)$, a weight connecting hidden neuron $x_l(k)$ to input $u_m(k)$. Similarly, each weight between the hidden and output layers is represented by $W_{nl}(k)$, a weight connecting output $\hat{y}_n(k)$ to hidden neuron $x_l(k)$. The weights are fully connected between layers. The desired outputs at iteration k are represented by $\mathbf{y}(k)$. Please note that common nomenclature for the total number of layers is based on the sum of the hidden and output layers. The input layer is excluded since no processing occurs. Therefore, the network in Fig. 2.20 contains two layers.

The processing at each hidden neuron may be represented as

$$x_l(k) = g\left[f_l(k)\right] = g\left[\sum_{m=1}^{M} w_{lm}(k)u_m(k)\right], \qquad (2.56)$$

where $g(x)$ is an activation function that is typically assumed to be the hyperbolic tangent (tanh) function times a constant of $1/2$. The derivative of this activation function is

$$g'(x) = 1 - g^2(x). \qquad (2.57)$$

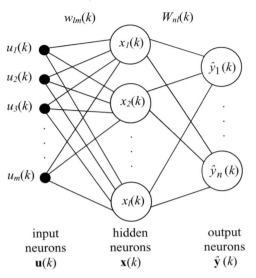

FIGURE 2.20: Multilayer feedforward network with two layers. An activation function, $g(x)$, and summation are contained within each neuron

The processing at each output may be represented as

$$\hat{y}_n(k) = g\left[f_n(k)\right] = g\left[\sum_{l=1}^{L} W_{nl}(k)x_l(k)\right] = g\left\{\sum_{l=1}^{L} W_{nl}(k)g\left[\sum_{m=1}^{M} w_{lm}(k)u_m(k)\right]\right\}. \quad (2.58)$$

Here, the thresholds have been omitted in these definitions, but could easily be added to the summation terms. Because this function is continuous and differentiable, system error may be propagated back from the outputs of the network to the inputs. This solution of "back propagation" was first described by Paul Werbos in 1974 (Werbos 1974), and was rediscovered independently by David Rumelhart (Rumelhart et al. 1986) and David Parker (Parker 1985).

2.3.2.1 Parameter Estimation

For the linear ARMAX model, we derived a performance function based on mean squared error. For the nonlinear multilayer feedforward network, let us derive a performance function based on squared error:

$$\xi[\mathbf{w}(k)] = \frac{1}{2}\sum_{n=1}^{N}\varepsilon^2(k) \quad (2.59)$$

$$\xi[\mathbf{w}(k)] = \frac{1}{2}\sum_{n=1}^{N} y_n[(k) - \hat{y}_n(k)]^2, \quad (2.60)$$

where $\mathbf{w}(k)$ contains weights from both layers, w_{lm} and W_{nl}. Substituting Eq. (2.58) into (2.60) yields

$$\xi[\mathbf{w}(k)] = \frac{1}{2} \sum_{n=1}^{N} \left\{ y_n(k) - g \left[\sum_{l=1}^{L} W_{nl}(k) g \left[\sum_{m=1}^{M} w_{lm}(k) u_m(k) \right] \right] \right\}^2. \qquad (2.61)$$

Since Eq. (2.61) is clearly a continuous differentiable function of every weight, we may use a gradient descent method to learn appropriate weights, as we did in adaptive filtering. With this method, we change each weight by an amount proportional to the gradient of the performance function at the present location. Let us calculate the partial derivatives of the performance function to obtain our learning rules.

For the hidden-to-output connections, the gradient descent method results in

$$\Delta W_{nl}(k) = \frac{\partial \xi[\mathbf{w}(k)]}{\partial W_{nl}(k)} = - \sum_{n=1}^{N} [y_n(k) - \hat{y}_n(k)] g'[f_n(k)] x_l(k). \qquad (2.62)$$

Similarly, for the input-to-hidden connections, the gradient descent method results in

$$\Delta w_{lm}(k) = \frac{\partial \xi[\mathbf{w}(k)]}{\partial w_{lm}(k)} = \frac{\partial \xi[\mathbf{w}(k)]}{\partial x_l(k)} \cdot \frac{\partial x_l(k)}{\partial w_{lm}(k)} \qquad (2.63)$$

$$\Delta w_{lm}(k) = - \sum_{n=1}^{N} \left\{ [y_n(k) - \hat{y}_n(k)] g'[f_n(k)] W_{nl}(k) \right\} \cdot g'[f_l(k)] u_m(k). \qquad (2.64)$$

We may then construct our learning rules as

$$W_{nl}(k+1) = W_{nl}(k) + \rho \Delta W_{nl}(k) \qquad (2.65)$$

$$w_{lm}(k+1) = w_{lm}(k) + \rho \Delta w_{lm}(k), \qquad (2.66)$$

where ρ is the learning rate.

Equations (2.62) and (2.64)–(2.66) constitute the back propagation algorithm. These equations back propagate the error within the system until it is minimized. To start, the weights are initialized to random values. During each iteration, k, of training, an input–output pair, or pattern, is presented to the network. As the weights are updated, the performance function decreases, adaptive to the local gradient. If the patterns are presented in random order, the path through the control space becomes stochastic, allowing wider exploration of the control surface. Since the update rules are local, minimal storage is required.

2.3.2.2 Model Validation

As previously stated, model validation involves analyzing the results of parameter identification to select an "optimum" model. For nonlinear models, we evaluate the correlation matrix and goodness of fit. Residual statistics are not applicable, since input data for neural networks may be presented in random order during training.

The correlation matrix of the hidden neurons, $\mathbf{R}_h(k)$, can be estimated as

$$\mathbf{R}_h(k) = \frac{1}{P} \sum_{i=1}^{P} \mathbf{x}_i(k)\mathbf{x}_i^T(k), \qquad (2.67)$$

where P is the number of input–output pairs and i is the subscript designating a particular pair. The necessary number of hidden neurons is determined by calculating the rank of this matrix. If a hidden neuron is expressed by a linear combination of other neurons, then this hidden neuron is redundant (Hu et al. 1991).

Goodness of fit may be determined using a more generalized version of Akaike's final prediction error criterion—the information criterion, AIC (Akaike 1974). AIC is applicable to nonlinear models. It is calculated as

$$\text{AIC} = NN \ln \sum_{i=1}^{P} \frac{1}{\sigma^2(k)} \left[y(k) - \hat{y}(k)\right]^2 + 2p, \qquad (2.68)$$

where $\sigma^2(k)$ is the system variance and p is the number of parameters.

2.3.2.3 Artificial Neural Network Example

As an example of a feedforward network, let us discuss the ability to recognize handwritten zip code from the U.S. mail. A network to solve this problem was developed by Le Cun et al. of AT&T Bell laboratories (Le Cun et al. 1989). The example database used for training and testing consisted of 9298 isolated numerals digitized from handwritten zip codes. Typical examples are shown in Fig. 2.21.

Note the large variety of sizes, writing styles, instruments, and writing quality. Many of the digits would be difficult for a human to classify. 7291 sample digits were used in training; 2007 digits were used in testing. Each digit was first normalized to fill an area consisting of 40×60 black and white pixels. These patterns were then reduced to 16×16 pixel images using a linear transformation that mapped the grey levels of the image into a range of $\{-1, +1\}$.

The network used possessed three hidden layers—J1, J2, and J3—and an output layer (Fig. 2.22). J1, the first hidden layer, consisted of 768 neurons. These neurons were arranged as 12 feature detectors, each composed of 64 neurons. Each group of 64 neurons was arranged in an 8×8 square; each neuron received information only from a 5×5 contiguous square of pixels of the original input. All 64 neurons within a feature detector possessed the same 25 weight values. The location of each 5×5 square shifted by 2 pixels between neighbors in the

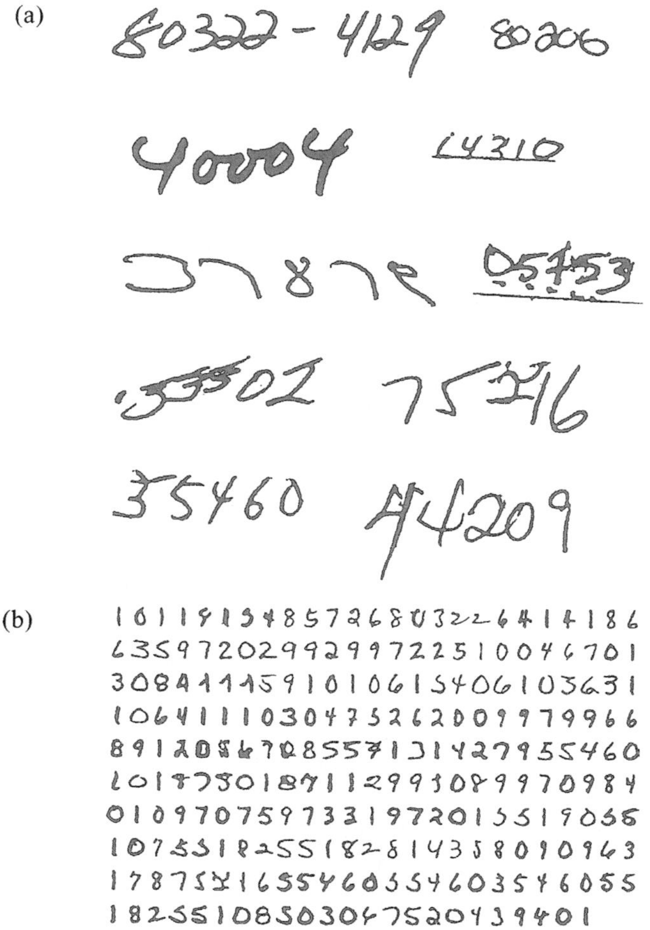

(a)

(b)

FIGURE 2.21: (a) Examples of handwritten zip codes and (b) normalized digits from the training/test database. Reprinted from Le Cun et al. (1989) with permission from MIT Press, Cambridge, MA

hidden layer. These extra arrangements enabled each feature detector, composed of 64 neurons, to detect one feature with high resolution.

Normally, all 256 inputs would be fully connected to all 768 neurons, requiring 196,608 weights. Using special weighting rules, the number of connections were reduced to 19,968 weights. As a result, only 768 threshold and 300 weights remained as free parameters in the first hidden layer during training.

The second hidden layer, J2, similarly consisted of 192 hidden neurons, arranged as 12 feature detectors. Each feature detector was composed if 16 (4 × 4) neurons each. Each neuron received information only from groups of 25 neurons, arranged as 5 × 5 receptive fields in the

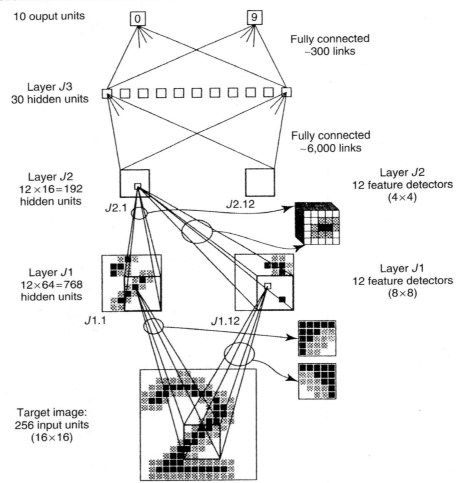

FIGURE 2.22: Architecture of multilayer feedforward network for handwritten character recognition. Reprinted from Le Cun et al. (1989) with permission of MIT Press, Cambridge, MA

previous layers. Using special weighting rules, the number of connections was reduced from 38,592 for full connection to 2592. As a result, only 192 thresholds and 200 weights remained as free parameters in the second hidden layer.

The third hidden layer, J3, consisted of 30 neurons, receiving information form all 192 neurons of the second layer. With full connections, J3 contained 5760 weights and 30 thresholds. The 10 output neurons, representing 10 digits, were fully connected to J3, and contained 300 weights and 10 thresholds.

This network was trained using back propagation. The training set was presented to the network 23 times in random order, and assumed that each digit occurred with the same probability in the training set. The weights and threshold were initialized with random values. The outputs were continuous values with the range $\{-1, +1\}$, rather than the typical ± 1 extreme values used in classification. This prevented the weights from growing indefinitely during training.

After training, only ten digits were misclassified (0.14%). However, 102 mistakes (5%) were reported during testing. By rejecting some of the more illegible test patterns (12.1%), the misclassification during testing was decreased to 1%. Simpler networks with fewer feature mapping levels were also evaluated, but produced inferior results (Le Cun et al. 1989).

2.3.3   Fuzzy Models and Control

For a complicated physiologic signal not easily represented by closed form equations, the system operator, $\mathbf{H}(k)$, in Fig. 2.2 may be described by a fuzzy model. This system operator is assumed to be nonlinear and time-invariant.

Fuzzy models are based on fuzzy sets, which were introduced by Dr. Lotfi Zadeh in 1965 (Zadeh 1965). Fuzzy sets are mathematical representations of the vagueness present in our natural language when we describe phenomena that do not possess sharply defined boundaries. When Zadeh published a journal article in 1973 describing the application of fuzzy logic to control (Zadeh 1973), Ebrahim Mamdani, then working on his dissertation at Queen Mary College in London, was inspired to create the first fuzzy controller (Mamdani and Assilian 1975).

While many types of fuzzy models now exist, we consider the model Mamdani originally created, which is known as the linguistic fuzzy model or Mamdani model (Fig. 2.23(a)). In this model, traditional crisp inputs are transformed into fuzzy inputs using fuzzification. Rule base inference is then used to map the fuzzy inputs into fuzzy outputs. The fuzzy outputs are transformed into crisp outputs by defuzzification. The knowledge base provides the fuzzy control rule base for fuzzy inference and data for fuzzification and defuzzification membership functions.

A fuzzy controller is merely the addition of feedback. Now, crisp outputs from the controller are fed back to a controlled system, with system outputs functioning as the crisp inputs to the controller (Fig. 2.23(b)). The simplest controller utilizes two crisp inputs and one crisp output. Often, these inputs are an error parameter and change in error parameter. As a rule of thumb, a minimum of five membership functions per input is required for smooth control.

Membership functions describe the relationships between crisp and fuzzy values. For example, a girl whose height is a crisp value of $5'5''$ may also be considered average in height (as opposed to very small, small, tall, or very tall). Given a fuzzy system with multiple crisp inputs, $\mathbf{u}(k)$, these crisp inputs are transformed into fuzzy inputs, $\mathbf{U}(k)$. Each fuzzy input, $U_{mi}(k)$ is an

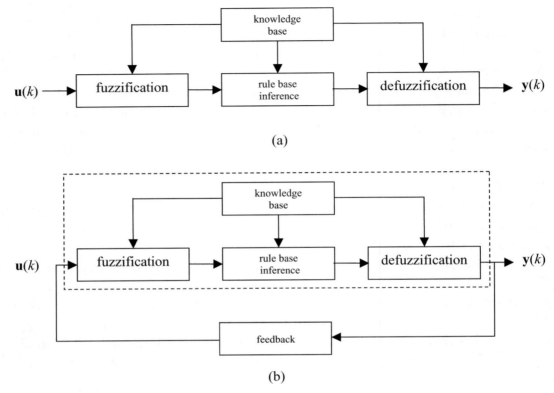

FIGURE 2.23: (a) Fuzzy model and (b) fuzzy controller

ordered pair containing a linguistic label, *label$_i$*, and degree of membership, $\mu_{mi}[u_m(k)]$:

$$U_{mi}(k) = \{label_i, \mu_{mi}[u_m(k)]\}, \hspace{3cm} (2.69)$$

where k refers to the sample and m refers to the input number. The degree of membership may also be referred to as the possibility (as opposed to statistical probability) with which a crisp input belongs to a membership function.

A family of membership functions with certain constraints is called a fuzzy partition. Typically, a fuzzy partition contains five to nine trapezoids. Note that a triangular function is a subset of a trapezoid. For any crisp input, all degrees of memberships sum to 1. All crisp inputs are assigned at least one membership function with a nonzero degree of membership (Fig. 2.24).

Once the fuzzy inputs are determined, the input labels are used to derive the output labels for various combinations, based on a given set of rules. The various combinations of fuzzy input degrees of membership are also combined using fuzzy logic to determine the output degrees of membership. This process is known as rule base inference. As with the fuzzy inputs,

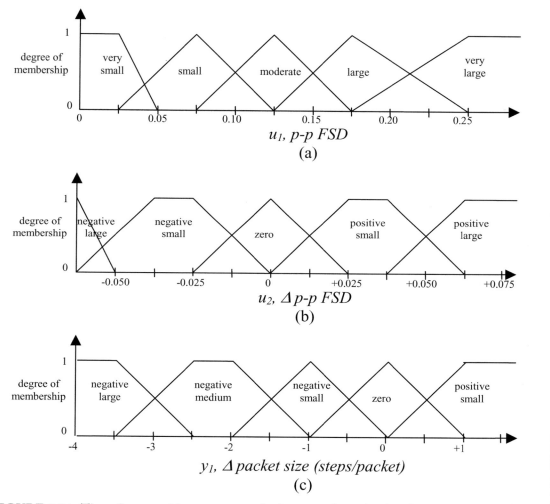

FIGURE 2.24: Three fuzzy partitions—composed of groups of membership functions. In the context of discussion, (a) and (b) are inputs, while (c) is an output

each fuzzy output, $Y_{ni}(k)$, consists of an ordered pair containing a linguistic label and degree of membership:

$$Y_{ni}(k) = \{label_i, \mu_{ni}[y_n(k)]\}. \tag{2.70}$$

Each predetermined rule, or fuzzy conditional statement, consists of an antecedent and a consequent. An antecedent contains several preconditions; a consequent contains one or more output actions. Both utilize linguistic labels. For example, a syringe pump motor controller rule may be "IF p-p FSD is large and Δp-p FSD is positive large, THEN Δ packet size is negative medium." In this example, p-p = peak-to-peak, FSD = fractional standard deviation, and Δ = change in. A second rule may be "IF p-p FSD is very large and Δp-p FSD is positive large, THEN Δ packet size is negative large." The maximum number of rules is equal to the product of the membership functions. For efficient implementation, rules are often arranged in tables.

Similarly, input degrees of membership are used to derive the output degrees of membership. Let us assume that the mathematical operation of taking the minimum, also known as the Zadeh intersection, is used to combine fuzzy input degrees of membership.

As an illustration of rule base inference, let us assume that the syringe pump motor controller sensors observe p-p FSD = 0.225 and Δp-p FSD = 0.0625, and that we can neglect k. From our input fuzzy partitions, p-p FSD = 0.225 transforms into $U_{11} = \{large, 0.33\}$ and $U_{12} = \{very\ large, 0.67\}$. Also, Δp-p FSD = 0.225 transforms into $U_{21} = \{positive\ large, 1\}$. Based on our stated rules and the Zadeh intersection, our fuzzy outputs are $Y_{11} = \{negative\ medium, 0.33\}$ and $Y_{12} = \{negative\ large, 0.67\}$.

In the last step of our process, we defuzzify the fuzzy outputs to crisp outputs. Defuzzification procedures are used to select an adequate decision among those deemed adequate by the output possibility distribution. Because one or more outputs may have been determined by rule base inference, these output sets are combined. According to the disjunctive interpretation of a fuzzy relation, which was posed by Zadeh, Mamdani, and Assilian, the outputs must be combined by union to approximate the compatibility relation. From the union of the outputs, the crisp output is then determined by taking the centroid (center of area):

$$y_{ni}(k) = \frac{\sum_{s=1}^{q} \mu_{ns}\left[y_s(k)\right] y_s(k)}{\sum_{s=1}^{q} \mu_{ns}\left[y_s(k)\right]}, \tag{2.71}$$

where $y_s(k)$ are the discrete points in the relevant domain.

For our syringe pump motor controller example, defuzzification is illustrated in Fig. 2.25. Here, $Y_{11} = \{negative\ medium, 0.33\}$ and $Y_{12} = \{negative\ large, 0.67\}$ have been combined, with the crisp output determined from the centroid calculation as $y_1 = -3.0$.

2.3.3.1 Parameter Estimation

The knowledge base comprises knowledge of the application domain and the modeling goals. From the knowledge base, the input and output fuzzy partitions and rule base are established. For traditional engineering systems, the characteristics of human control behavior, development of process skills, individual differences between process operators, task factors affecting performance, and organization of the operator's behavior must be taken into account during construction of the fuzzy partitions and rule base. For physiologic systems, natural variation must also be generalized from sufficient observations of the control space. The membership functions are selected to serve as meanings for the linguistic labels in the inference rules.

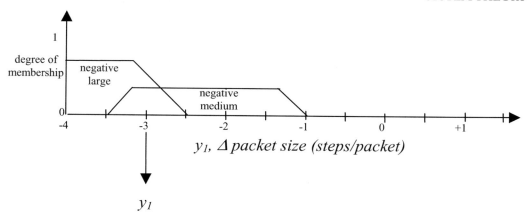

FIGURE 2.25: Syringe pump motor controller example defuzzification

Typically, for a system designed exclusively on the basis of intuitive knowledge, fine tuning of the fuzzy partitions and rule base amounts to observing the results and then intuitively adjusting membership functions that were derived from intuition in the first place. For simple fuzzy models, such as those with two inputs, this trial and error method is adequate. However, with larger systems, a methodology such as neural networks may be utilized for efficient fine tuning (Keller and Tahani 1992).

2.3.3.2 Model Validation

Model validation is not as straightforward with a fuzzy model as with a traditional mathematical model. It is possible, however, to analyze the coverage of the input space by the rules. For an incomplete rule base, additional rules may be constructed from prior knowledge. The antecedents of these rules may be created from unused combinations of membership functions in the initial model. The identification data usually cover only a fraction of the complete product space of the model parameters. Therefore, the antecedents of the obtained rules include only those combinations of the linguistic terms that were identified from the data. It is possible that regions not covered by any rules are entered during simulation or prediction. This situation can be detected by observing the output degree of membership in the rule base. If no rule is activated above a specified threshold, an additional rule may be added to the rule base. The antecedent of this rule is given by the combination of linguistic terms that give the highest output degree of membership for the given data point (Babuska 1998).

2.3.3.3 Fuzzy Control Example

In 1963, Pressman and Newgard at the Stanford Research Institute postulated that the technique of ocular tonometry could be adapted to blood pressure. With ocular tonometry, a sensor is applied to the cornea until its central area is flattened. This flattening indicates that the circumferential stresses in the corneal wall have been removed and that the internal and external pressures are equal. The final pressure with which this sensor is applied equals the intraocular pressure. Similarly, Pressman and Newgard believed that sufficient pressure could be applied to an artery such as the radial artery, which possesses sufficient bony support. With sufficient pressure, the transmural (through the arterial wall) pressure would equal zero and the external pressure would equal the internal pressure (Pressman and Newgard 1963).

Several groups have investigated the application of arterial tonometry for continuous, noninvasive blood pressure monitoring since publication of this article. In the late 1990s, Baura developed a method for estimation of systolic, diastolic, and mean arterial pressure, based on time–frequency analysis of ultrasound velocity waveforms obtained at the radial artery. The estimates were obtained during an initial applanation sweep (data acquisition from overcompression to undercompression of the artery), as shown in Fig. 2.26. Once initial estimates were made, continuous measurement could be accomplished using fuzzy control (Baura 2002a, 2003).

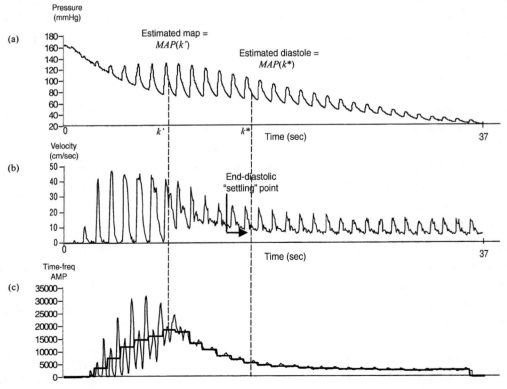

FIGURE 2.26: Blood pressure estimation during an applanation sweep. Based on (Baura 2003)

The fuzzy controller was based on the observation that the blood pressure beat possessing the true (intraarterial) mean arterial pressure (MAP) corresponds to the beat with the maximum end-diastolic blood velocity. The controller was implemented with two inputs: (1) the mean pseudo-Wigner distribution of the current beat, and (2) the difference between the current and last mean pseudo-Wigner distribution. The controller output was the number of applanation steps sent to the motor. This output ranged from −400 to +400 steps, in multiples of 50 steps. For the motor, 38,400 steps equaled 1 in. If the difference input was a positive value, the output signal directed the applanation motor to continue in the same direction for the calculated number of steps. If the difference input was a negative value, the output signal directed the applanation motor to change direction for a calculated number of steps. The input and output membership functions of the controller were typical functions of overlapping trapezoids. Unfortunately, these fuzzy partitions were not publicly disclosed.

In an initial trial in one surgical patient, the controller was evaluated during an observed 50-mmHg drop in MAP over approximately 11 min. This severe pressure drop occurred in response to epidural administration of the anesthetic bupivicaine. The controller tracked 552 pressure beats, with a mean difference of 3 ± 4 mmHg (Fig. 2.27). Figure 2.28 is a detail view

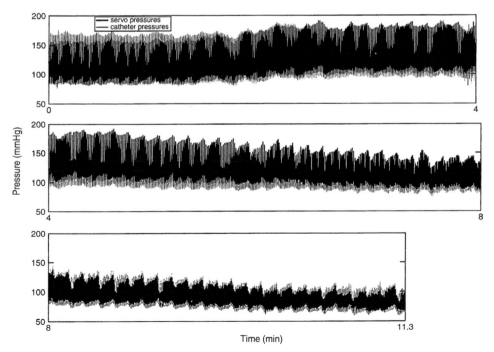

FIGURE 2.27: Continuous intraarterial catheter and servo pressures in a surgical patient over 11.3 min. During this time, the catheter MAP dropped 50 mmHg. Based on (Baura 2003)

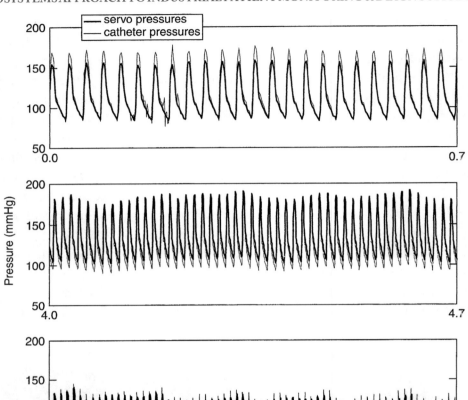

FIGURE 2.28: Forty-second snapshots of the catheter and servo data in Fig. 2.27. Based on (Baura 2003)

of Fig. 2.27, illustrating three 40-s windows of the intraarterial catheter and servo data. Figure 2.29 is a 20-s interval snapshot of the data in Fig. 2.28 that occurred at 6.5 min from the onset of data recording in Fig. 2.27. As illustrated in Fig. 2.29, a significant drop in the end-diastolic velocity was corrected within five beats.

The controller was then tested in two anesthetized surgical patients during two continuous 20-min intervals, with each test followed by a 5-min resting interval. During each 20-min measurement, one applanation pressure sweep was conducted, followed by continuous servo control. During the four measurement periods, the catheter MAPs ranged from 69 to 106 mm

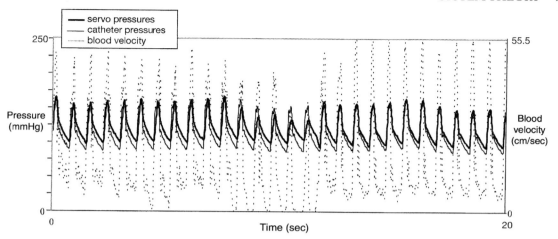

FIGURE 2.29: Twenty-second snapshot of catheter and servo data at 6.5 min. During this time, a significant drop in the end-diastolic velocity was corrected within five beats. Based on (Baura 2003)

Hg. Over 3103 beats, the mean MAP difference was -3 ± 5 mmHg. In data from all three surgical patients, the AAMI standard of $\leq 5 \pm 8$ mmHg was met for MAP differences. This work was discontinued when the inventor left VitalWave Corporation, now known as Tensys Medical.

REFERENCES

Akaike, H., "Statistical predictor identification," *Ann Inst Stat Math*, Vol. 22, pp. 203–217, 1970.

—— "A new look at the statistical model identification," *IEEE Trans Auto Cont*, AC-19, pp. 716–723, 1974.

Baura, G. D., "Method and apparatus for electrode and transthoracic impedance estimation," U.S. Patent 6,016,445, January 18, 2000a.

—— "Method and apparatus for high current electrode, transthoracic, and transmyocardial impedance estimation," U.S. Patent 6,058,325, May 2, 2000b.

—— "Method and apparatus for high current electrode, transthoracic, and transmyocardial impedance estimation," U.S. Patent 6,253,103, June 26, 2001.

—— "Method and apparatus for the noninvasive determination of arterial blood pressure," U.S. Patent 6,471,655, October 29, 2002a.

—— "Method and apparatus for the noninvasive determination of arterial blood pressure," U.S. Patent, 6,514,211, February 4, 2003.

—— *System Theory and Practical Application of Biomedical Signals*. Hoboken, NJ: Wiley-IEEE Press, 2002b.

Babuska, R., *Fuzzy Modeling for Control*. Boston: Kluwer, 1998.

Cybenko, G., "Approximation by superpositions of a sigmoidal function," *Mathematics of Control, Signals, and Systems*, Vol. 2, pp. 303–314, 1989.

Dahl, C. F., Ewy, G. A., Ewy, M. D., and Thomas, E. D., "Transthoracic impedance to direct current discharge: effect of repeated countershocks," *Med Instr*, Vol. 10, pp. 151–154, 1976.

Hornik, K., Stinchcombe, M., and White, H., "Multilayer feedforward networks are universal approximators," *Neural Networks*, Vol. 2, pp. 359–366, 1989.

Hu, Y. H., Xue, Q., and Tompkins, W. J. "Structural simplification of a feed-forward multi-layer perceptron artificial neural network," in *Proc. Int. Conf. on Acoustics, Speech, and Signal Proc.*, Vol. 2. Toronto, Canada, May 1991, pp. 1061–1064.

Kadambe, S., Murray, R., and Boudreaux-Bartels, G. F., "Wavelet transform-based QRS complex detector," *IEEE Trans BME*, Vol. 46, pp. 838–848, 1999.

Keller, J. M. and Tahani, H., "Backpropagation neural networks for fuzzy logic," *Info Sci*, Vol. 62, pp. 205–221, 1992.

Le Cun, Y., Boser, B., Denker, J. S., Henderson, D., Howard, R. E., Hubbard, W., and Jackel, L. D., "Backpropagation applied to handwritten zip code recognition," *Neural Comp*, Vol. 1, pp. 541–551, 1989.

Mamdani, E. H. and Assilian, S., "An experiment within linguistic synthesis with a fuzzy logic controller," *Int J Man-Machine Studies*, Vol. 7, pp. 1–13, 1975.

Murray, R., Kadambe, S., and Boudreaux-Bartels, G. F., "Extensive analysis of a QRS detector based on the dyadic wavelet transform," in *Proc. IEEE-SP Int. Symp. on Time-Frequency and Time-Scale Analysis*, Philadelphia, PA, 1994, pp. 540–543.

Oppenheim, A. V., Schafer, R. W., and Buck J. R. *Discrete-time Signal Processing*, 2nd ed., Englewood Cliffs, NJ: Prentice Hall, 1999.

Parker, D. B., *Learning Logic*. Technical Report TR-47, Center for Computational Research in Economics and Management Science, Cambridge, MA: MIT, 1985.

Pressman, G. L. and Newgard, P. M., "A transducer for continuous external measurement of arterial blood pressure," *IEEE Trans Biomed Elect*, Vol. 10, pp. 73–81, 1963.

Rumelhart, D. E., Hinton, G. E., and Williams, R. J., "Learning representations by back-propagating errors," *Nature*, Vol. 323, pp. 533–536, 1986.

Sugimachi, M. National Cardiovascular Center Research Institute, Osaka, Japan. Research data, 1992.

Voss, G. I., Butterfield, R. D., Baura, G. D., and Barnes, C. W. "Fluid flow impedance monitoring system," U.S. Patent 5,609,576, March 11, 1997.

Werbos, P., "Beyond regressions: new tools for prediction and analysis in the behavioral sciences," Ph.D. thesis, Boston, MA: Harvard University, 1974.

Widrow, B. and Hoff, M., Jr., "Adaptive switching circuits," *IRE WESCON Conv Rec*, Part 4, pp. 96–104, 1960.

Widrow, B., Glover, J. R., McCool, J. M., Kaunitz, J., Williams, C. S., Hearn, R. H., Zeidler, J. R., Dong, E., and Goodlin, R. C., "Adaptive noise canceling: principles and applications," *Proc. IEEE*, Vol. 63, pp. 1692–1716, 1975.

Widrow, B. and Sterns, S. D., *Adaptive Signal Processing*. Englewood Cliffs, NJ: Prentice Hall, 1985.

Zadeh, L. A., "Fuzzy sets," *Information and Control*, Vol. 8, pp. 338–353, 1965.

—— "Outline of a new approach to the analysis of complex systems and decision processes," *IEEE Trans SMC*, Vol. 3, pp. 28–44, 1973.

CHAPTER 3

Patient Monitoring Devices

In this chapter, we describe four patient monitoring devices that use system theory to minimize the noise in their physiologic input signals. Since patient monitoring involves continuous observation and processing, a high signal-to-noise ratio is a critical first step to obtaining accurate device parameters.

Masimo used adaptive filtering to minimize motion artifact during pulse oximetry measurements. Interflo Medical used the pseudorandom binary sequence to increase the signal-to-noise ratio during continuous thermodilution measurements. CardioDynamics used wavelet transforms to minimize the noise artifact during impedance cardiography measurements. Aspect Medical Systems used black box system identification to identify the stages of anesthesia administration.

3.1 MASIMO PULSE OXIMETRY

Masimo Corporation was founded in 1989 by electrical engineers Joe Kiani and Mohamed Diab, with the intent of minimizing motion artifact during pulse oximetry using adaptive filtering. Pulse oximetry measures the arterial saturation of oxygen, S_aO_2, which is the percentage of oxygen bound to hemoglobin in arterial blood. Arterial saturation of oxygen is related to the partial pressure of oxygen, P_{O2}, which is a critical operating room parameter that is monitored during mechanical ventilation to ensure oxygen is reaching the tissues. S_pO_2 represents an S_aO_2 measurement using pulse oximetry.

During the late 1980s and early 1990s, pulse oximetry began to gain acceptance by the medical community and migrated from use in the operating room to other hospital units. The American Society of Anesthesiology adopted pulse oximetry as a standard of basic care in the operating room in 1990 and in the recovery room in 1992 (American Society of Anesthesiologists 1991). As pulse oximetry measurements became more widespread, measurements were made in less controlled environments in which patient motion was present. Masimo believed adaptive filtering could be used to minimize noise in the face of motion and poor peripheral perfusion (low signal-to-noise ratio during low blood volume).

3.1.1 Pulse Oximetry Calculations

Theoretically, pulse oximetry is based on the Beer–Lambert law. As described by this law, the intensity of transmitted light, I, traveling through a uniform medium of path length, l, containing an absorbing substance with concentration, c, decreases as

$$I = I_0\, e^{-\varepsilon(\lambda)cl}, \tag{3.1}$$

where I_0 is the intensity of the incident light, λ is a specific wavelength, and ε is the extinction coefficient of the absorbing substance at that wavelength. The unscattered absorbance, $A(\lambda)$, of this process is the natural log of the ratio of transmitted light to incident light intensity:

$$A(\lambda) = -\ln \frac{I}{I_0} = \varepsilon(\lambda)cl. \tag{3.2}$$

When more than one substance absorbs light in a medium, each absorber contributes its part to the total absorbance, $A_T(\lambda)$, as:

$$A_T(\lambda) = \sum_{i=1}^{n} \varepsilon_i(\lambda)c_i l_i, \tag{3.3}$$

where n equals the number of absorbers.

If two wavelengths of light are transmitted through a glass cuvette containing a blood sample containing only oxyhemoglobin (hemoglobin bound to oxygen, HbO_2) and deoxyhemoglobin, Hb, the arterial saturation of oxygen can be determined as

$$S_aO_2 = \frac{[HbO_2]}{[Hb] + [HbO_2]}, \tag{3.4}$$

where [] denotes concentration. Please note that in special, rare cases of anemic hypoxia and carbon monoxide poisoning, two other forms of hemoglobin may be found in an arterial blood sample. Because the extinction coefficients of oxyhemoglobin and deoxyhemoglobin differ at each wavelength (Fig. 3.1), their respective concentrations can be found from Eq. (3.3). The concentrations are then input into Eq. (3.4) to determine S_aO_2.

In reality, transmitting light through a glass cuvette is much simpler than transmitting light through a finger or an ear lobe. Ludwig Nicolai began his oximetry experiments in 1931. Forty-one years later, pulse oximetry could only be used to accurately measure S_aO_2 if the S_aO_2 measurement was at least 90%, which is within the range of healthy individuals. This limited measurement range was enabled by Takuo Aoyagi at Nihon Kohden in 1972, who realized that only the pulsating components (arterial and venous blood) should be considered during measurement. This realization enabled other primary absorbers such as skin pigmentation and bones, and nonspecific sources of optical attenuation, to be discounted. For a complete derivation, please see Baura (2002).

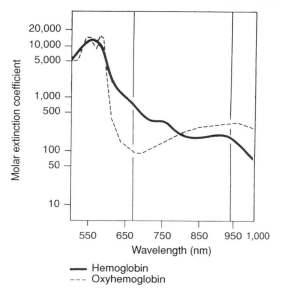

FIGURE 3.1: The extinction coefficients of deoxyhemoglobin (or hemoglobin) and oxyhemoglobin as functions of wavelength. Wavelengths employed by the pulse oximeter (660 and 940 nm) are indicated. Reprinted from (Mackenzie, 1985) with kind permission of Spring Science and Business Media.

The parameter that Aoyagi used to estimate S_pO_2 was the ratio, R, which is the ratio of the first time derivative of total absorbances at two wavelengths. This equation is given below in discrete time, after the time derivative for path length has been cancelled out of the numerator and denominator:

$$R(k) = \frac{\varepsilon_o(\lambda_1)c_o(k) + \varepsilon_d(\lambda_1)c_d(k)}{\varepsilon_o(\lambda_2)c_o(k) + \varepsilon_d(\lambda_2)c_d(k)}, \qquad (3.5)$$

where k is a discrete sample, the subscript "o" represents oxyhemoglobin, and the subscript 'd' represents deoxyhemoglobin. $R(k)$ was used to calculate S_pO_2 as:

$$S_pO_2(k) = \frac{\varepsilon_d(\lambda_1) - \varepsilon_d(\lambda_2)R(k)}{[\varepsilon_d(\lambda_1) - \varepsilon_o(\lambda_1)] - [\varepsilon_d(\lambda_2) - \varepsilon_o(\lambda_2)]R(k)}. \qquad (3.6)$$

Since these calculations did not hold for $S_aO_2 < 90\%$, Scott Wilber, who founded Biox Technology, instead used calibration curves to derive the relationship between $R(k)$ and S_pO_2. Specifically, he measured constant $R(k)$ in healthy volunteers who were inspiring a specific oxygen concentration. The resulting calibration curve (Fig. 3.2) provided a simple method for determining arterial oxygen saturation. This calibration curve enables accurate

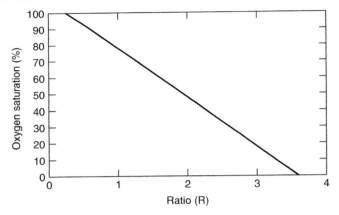

FIGURE 3.2: Typical pulse oximetry calibration curve

pulse oximetry measurements when motion artifact is not present. Please note that each pulse oximeter monitor/sensor combination requires its own calibration curve.

3.1.2   Adaptive Filtering in Masimo Software

Masimo used the signal and noise in two wavelengths of light from pulse oximetry to essentially derive and filter out the noise. In classic adaptive filtering, a noisy signal and a reference noise source are available for processing. In Masimo's adaptive noise cancellation application, a relationship for the reference noise source is derived.

The ratio in Eq. (3.5) can be rewritten to describe the ratio of the input signal, $u(k)$, at two wavelengths, λ_1 and λ_2:

$$R(k) = \frac{\varepsilon_o(\lambda_1)c_o(k) + \varepsilon_d(\lambda_1)c_d(k)}{\varepsilon_o(\lambda_2)c_o(k) + \varepsilon_d(\lambda_2)c_d(k)} = \frac{u_{\lambda_1}(k)}{u_{\lambda_2}(k)}. \qquad (3.7)$$

In the original derivation, it was assumed that noise was not present. Let us define the signals from the photodiodes, the primary signals $y_{\lambda_i}(k)$, as the combination of the input and noise, $n_{\lambda_i}(k)$, such that

$$y_{\lambda_1}(k) = u_{\lambda_1}(k) + n_{\lambda_1}(k) \qquad (3.8)$$

$$y_{\lambda_2}(k) = u_{\lambda_2}(k) + n_{\lambda_2}(k). \qquad (3.9)$$

Rearranging Eqs. (3.8) and (3.9) and substituting them into Eq. (3.7) yields

$$R(k) = \frac{y_{\lambda_1}(k) - n_{\lambda_1}(k)}{y_{\lambda_2}(k) - n_{\lambda_2}(k)}. \qquad (3.10)$$

Cross-multiplication and rearrangement results in

$$R(k)y_{\lambda_2}(k) - R(k)n_{\lambda_2}(k) = y_{\lambda_1}(k) - n_{\lambda_1}(k) \tag{3.11}$$

$$n_{\lambda_1}(k) - R(k)y_{\lambda_2}(k) = y_{\lambda_1}(k) - R(k)y_{\lambda_2}(k) \equiv n_1(k). \tag{3.12}$$

Note that we have defined the reference noise source, $n_1(k)$, as the linear combination of the detected noise source and the ratio. This reference noise source will effectively vary with motion artifact and poor perfusion. However, without noise, $n_1(k)$ equals zero and

$$R(k) = \frac{y_{\lambda_1}(k)}{y_{\lambda_2}(k)}. \tag{3.13}$$

Unfortunately, the reference noise source depends on the ratio, $R(k)$, which is the parameter being monitored to calculate S_pO_2. Since we have two unknowns but only one equation, 117 possible values of the ratio which correspond to uniformly spaced S_pO_2 values from 34.8 to 105% are used to calculate candidate reference noise source. Each candidate reference noise source is then input to the adaptive noise canceller with the infrared primary input; a corresponding filter output is determined. It is assumed by Masimo that the peak of the output power at the highest saturation corresponds to the arterial saturation (Fig. 3.3).

This sequence of saturation calculations is repeated once per second (Diab et al. 1997). The system configuration is given in Fig. 3.4; a Masimo Radical-7 monitor that incorporates

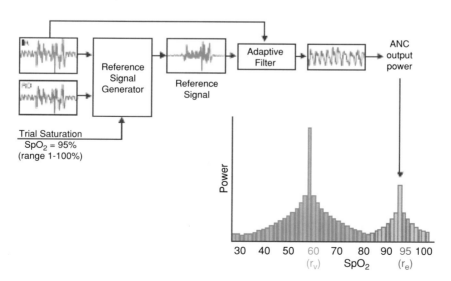

FIGURE 3.3: S_pO_2 selection, based on peak with highest saturation. ANC = adaptive noise cancellation. Courtesy of Masimo Corporation, Irvine, CA

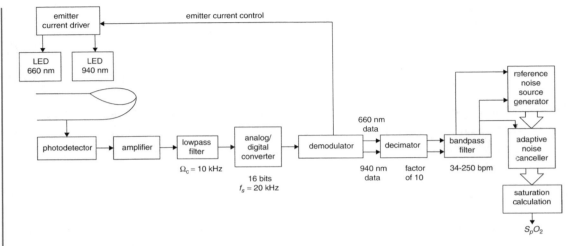

FIGURE 3.4: Masimo system configuration. Based on (Diab et al. 1997)

adaptive noise cancellation (trademarked Signal Extraction Technology or SET) is given in Fig. 3.5.

3.1.3   Clinical Results

This system for motion resistant pulse oximetry was validated in 10 healthy volunteers. Each volunteer was monitored using three different pulse oximeters: the Nellcor N-200, with its heuristic & inaccurate C-LOCK algorithm not used; the Nellcor N-3000 that used Oxismart "advanced signal processing"; and a Masimo prototype that used adaptive noise cancellation. Masimo chose Nellcor monitors for comparison testing, as Nellcor was the market leader. A

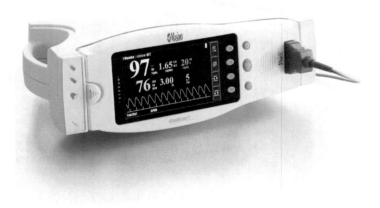

FIGURE 3.5: Masimo Radical-7 monitor. Courtesy of Masimo Corporation, Irvine, CA.

TABLE 3.1: Pulse oximeter performance index (*PI*, %) during motion. Based on (Barker 1997)

	NELLCOR N-200	NELLCOR N-3000	MASIMO PROTOTYPE
Oximeter connected before motion begins	76	87	99
Oximeter connected after motion begins	68	47	97
Oximetry readings at lowest point of rapid desaturation to 75%	70	58	95

disposable sensor corresponding to each monitor was randomly placed on the index, middle, or ring finger of each subject's dominant hand. A sensor for a reference pulse oximeter was also placed on the nondominant hand. Each subject inspired various oxygen fractions from an anesthesia machine through a tight-fitting face mask to simulate room air, steady-state hypoxemia ($S_pO_2 \approx 75\%$), and transient hypoxemia (S_pO_2 varied between 75 and 100%). During each oxygen state, the dominant hand with sensors was subjected to standardized "rubbing" and "tapping" motions generated by a motor-driven tilt table. Further, the sensors were disconnected and reconnected at various preselected times, forcing the instrument to reacquire data during the motion conditions.

The Masimo pulse oximeter performed significantly better ($p < 0.05$) than the other monitors during motion in terms of performance index, *PI*, as shown in Table 3.1. In this study, *PI* was defined as the time percentage during which the oximeter provided an S_pO_2 value within 7% of the control S_pO_2 value (Barker 1997). Typical S_pO_2 readings during rapid desaturation-resaturation and tapping motion are shown in Fig. 3.6.

3.1.4   Conclusion

In 1999, Masimo sued Nellor for patent infringement when Nellcor released its N-395 pulse oximeter, which Nellcor claimed could monitor in the presence of patient motion. Nellcor became part of the healthcare division of Tyco International in 2000. In 2007, Tyco's healthcare division was spun off as Covidien.

In March, 2004, a jury found that Nellcor infringed several Masimo patents. In September 2005, the appellate court affirmed the infringement findings against Nellcor, and instructed the District Court to enter a permanent injunction against Nellcors' pulse oximeters that were found to infringe (N-395 and N-595). In January 2006, Masimo and Nellcor entered into a

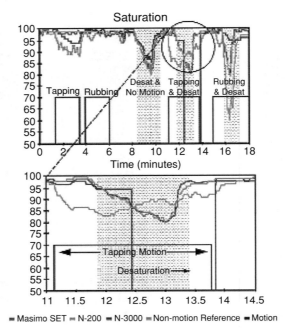

FIGURE 3.6: S_pO_2 versus time showing a rapid desaturation-resaturation occurring during tapping motion. Courtesy of Masimo Corporation, Irvine, CA

settlement agreement, where Nellcor agreed to discontinue shipment of all pulse oximeters that were found to infringe Masimos' patents and pay Masimo $265 million for sales of the infringing products (Masimo Corporation 2006). This was a momentous patent infringement decision for the medical device industry, as it demonstrated that innovative startups could prevail over large conglomerates.

In 1997, hospital sales of pulse oximeters and their disposable probes generated $32,011,546 and $197,494,350, respectively, in the United States. These sales estimates exclude sales in federal hospitals and nursing homes. Nellcor accounted for 88.1% of total disposable sales (IMS Health 1997, Baura 2002). By 2006, nonfederal hospital sales of pulse oximeters and their disposable probes were $11,766,191 and $430,426,031 respectively. Nellcor accounted for 74.9% of total disposable sales (IMS Health 2006). With the settlement of the Masimo v. Nellcor patent infringement lawsuit, it is believed that Masimo sales will significantly increase in the future.

3.2   INTERFLO MEDICAL CONTINUOUS THERMODILUTION

Interflo Medical was founded in the late 1980s by anesthesiologist and electrical engineer Dr. Mark Yelderman, with the intent of converting thermodilution from an intermittent to continuous measurement. Yelderman originally conceived of this invention while still a physician at Stanford University (Yelderman 1985). Thermodilution measures cardiac output, CO, which is the effective blood volume expelled by either ventricle per unit time. Typically, it is calculated as the product of the left ventricular stroke volume times the heart rate. Traditionally, cardiac output was considered necessary to guide therapy for certain critically ill patients, such as those who had experienced complicated cardiac surgery, complicated mechanical ventilation, emergency or extensive surgery, acute pulmonary edema, or acute lung injury. It was generally believed that CO management led to better patient outcomes.

Thermodilution is one of a group of indicator-dilution methods, by which a detectable indicator is applied upstream in the circulation and detected downstream to determine the flow rate by which it was mixed. It is assumed that the indicator mixes with all the blood flowing through the central mixing pool. The indicator for thermodilution is a bolus of room temperature or iced saline. Because of the nature of this measurement, thermodilution can only be made intermittently. Interflo Medical believed that the iced saline could be replaced by heating of the blood itself. Since it is required that the blood should not be affected by significantly increased temperature, small pulses of heat pulses were applied upstream. To increase the signal-to-noise ratio of these pulses, the pulses were coded as a pseudorandom binary sequence.

3.2.1 Thermodilution Calculations

In preparation for thermodilution, a multiple lumen pulmonary artery catheter (PAC) is passed through the skin into a central vein. Once the catheter tip reaches a central venous location, a balloon at the tip is inflated, which causes the catheter tip to rapidly move from the right atrium, through the right ventricle, and into the pulmonary artery. A bolus of room temperature or iced (0°C) 5% dextrose in water or 0.9% NaCl is injected through the catheter into the right atrium. The volume injected in adult patients, that are not fluid restricted, is generally 10 ml. The resulting blood temperature transient is detected downstream by a thermistor in the pulmonary artery. A typical thermodilution curve is shown in Fig. 3.7.

Cardiac output is then calculated using the Stewart–Hamilton equation as

$$CO = \frac{V_I(T_B - T_I)K_1K_2}{A},$$

(3.14)

where V_I is the injectate volume, T_B is the blood temperature in the pulmonary artery, T_I is the injectate temperature, K_1 is the density factor (injectate/blood), K_2 is the catheter manufacturer's computation constant, and A is the area under the thermodilution curve. Due to heat loss through the catheter wall, several serial injections are needed to obtain a

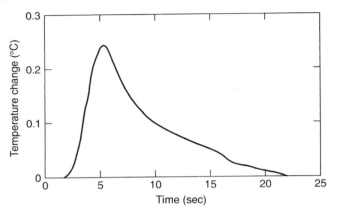

FIGURE 3.7: Typical intermittent thermodilution curve

consistent value for cardiac output. If the cardiac output is low, the resulting curve will be very broad, decreasing the signal-to-noise ratio of the measurement. Respiratory-induced variations in pulmonary artery blood temperature may confound the dilution curve when it is of low amplitude. Because of fluid overloading, this in vivo measurement can only be obtained intermittently.

3.2.2   PRBS in Interflo Medical Software

Interflo Medical designed a special PAC that had been modified to include a filamentous heating element (Fig. 3.8(a)). When the filament was positioned in the right ventricle, a distal thermistor was now positioned in the pulmonary artery to monitor temperature changes. A maximum of 15 W of heat was delivered to filament, coded as a pseudorandom binary sequence of length $M = 31$. This produced transient increased temperatures of $+0.02°C$ in the pulmonary artery (Fig. 3.9).

Using Eq. (2.9), the detected thermistor signal was decoded to obtain an amplified thermodilution curve, with an increased signal-to-noise ratio. A typical curve is shown in Fig. 3.10. The system configuration for these measurements is given in Fig. 3.11.

3.2.3   Clinical Results

This system for continuous thermodilution was validated first in sheep (Yelderman et al. 1992a) and then in humans. In the human studies, 54 intensive care unit (ICU) patients, ranging in weight from 54 to 111 kg, were each studied for six hours. Patient CO was determined by both continuous and bolus thermodilution, resulting in 222 data pairs. The bolus COs ranged from

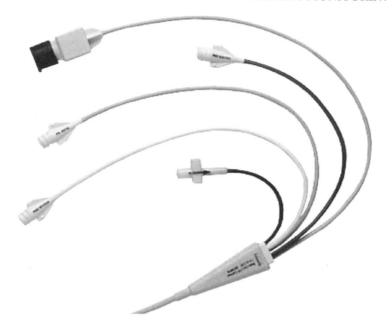

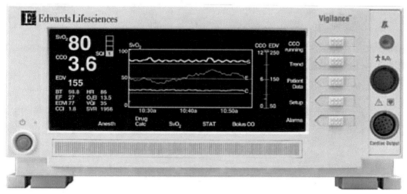

FIGURE 3.8: (a) Continuous thermodilution CCOMBO catheter. (b) Continuous thermodilution Vigilance monitor. Courtesy of Edwards Lifesciences, Irvine, CA

2.8 to 10.8 l/min (Fig. 3.12); the heart rates varied from 74 to 158 bpm, with some periods of irregular rhythms.

Using linear regression analysis, a squared correlation coefficient of $r^2 = 0.88$ (p not reported) was obtained, meaning that 88% of the variance in the continuous thermodilution measurements could be accounted for by the bolus thermodilution measurements. The absolute measure bias was 0.02 l/min; the 95% confidence limits were 1.07 and -1.03 l/min (Yelderman et al. 1992b).

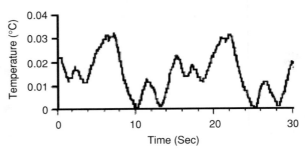

FIGURE 3.9: Thermistor temperature response. Reprinted from (Yelderman 1990) with kind permission of Springer Science and Business Media

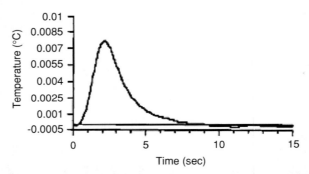

FIGURE 3.10: Typical continuous thermodilution curve. Reprinted from (Yelderman, 1990) with kind permission of Spring Science and Business Media.

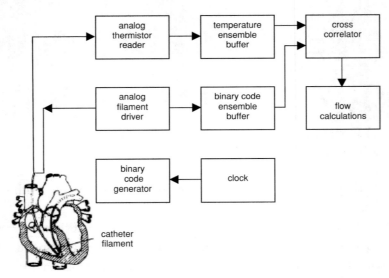

FIGURE 3.11: Interflo Medical system configuration. Reprinted from (Yelderman, 1990) with kind permission of Spring Science and Business Media.

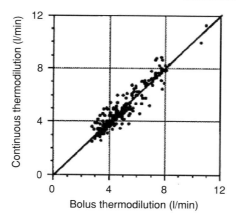

FIGURE 3.12: Continuous vs. bolus thermodilution measurements for 54 patients, 222 data pairs Reprinted from (Yelderman, 1992) with kind permission of Spring Science and Business Media.

3.2.4   Conclusion

Interflo Medical was acquired by Baxter Edwards in 1992. In 2000, Baxter Edwards spun off its cardiovascular group, which is now known as Edwards Lifesciences. Baxter Edwards marketed Interflo's continuous thermodilution system, without significant design changes, as the CCOMBO catheter and Vigilance monitor (Figs. 3.8(a) and (b)).

The market for both bolus and continuous thermodilution changed dramatically with the publication of the observational study of PAC use by Connors et al. in 1996. In this study of 5735 critically ill adult patients receiving ICU care in one of five U.S. teaching hospitals, the outcomes of patients receiving right heart catheterization within 24 h of ICU stay (38%) were compared to those who did not receive this initial treatment. By case-matching analysis, it was determined that PAC patients had an increased 30 day mortality and increased 2 day stay in the ICU (Connors et al. 1996).

Over time, this study caused PACs, and therefore thermodilution CO measurements, to decline in use. In 1997, hospital sales of thermodilution catheters and their monitoring accessories generated $67,162,910 and $14,604,868, respectively, in the United States. These sales estimates exclude sales in federal hospitals and nursing homes. Baxter accounted for 54.9 and 58.8% of total catheter and accessories sales, respectively (IMS Health 1997, Baura 2002). By 2006, nonfederal hospital sales of thermodilution catheters and their monitoring accessories had declined to $63,112,571 and $6,941,915 respectively (IMS Health 2006).

3.3   CARDIODYNAMICS IMPEDANCE CARDIOGRAPHY

CardioDynamics International Corporation traces its roots to BoMed Medical Manufacturing, which was founded by electrical engineer Bo Sramek in the 1980s. Sramek founded his company to commercialize monitors based on impedance cardiography. BoMed went through bankruptcy in 1992–3, and reemerged through reorganization as CardioDynamics.

Impedance cardiography is a noninvasive, continuous method for monitoring cardiac output. As discussed earlier in this chapter, CO is the effective blood volume expelled by either ventricle per unit time. Typically, it is calculated as the product of the left ventricular stroke volume times the heart rate. After CO measurements by invasive thermodilution were demonstrated to increase mortality in 1996 (Connors et al. 1996), it was believed by some that physicians would respond by increasing their use of noninvasive CO measurements.

3.3.1   Impedance Cardiography Calculations

Strictly speaking, impedance cardiography, which is also known as thoracic bioimpedance or impedance plethysmography, is used to measure cardiac stroke volume, SV. Stroke volume is the ejection volume from the left ventricle during systole. When the stroke volume is multiplied by heart rate, cardiac output is obtained:

$$CO = SV \cdot \text{heart\_rate}. \tag{3.15}$$

The original stroke volume measurement, based on thoracic impedance, was invented by Kubicek et al. at the University of Minnesota for use by NASA. Two band electrodes are positioned on the neck, with the inner electrodes at the root of the neck. Two band electrodes are positioned at edge of the thorax, with the inner electrode at the level of the xyphoid process. Constant current flows through the outer electrodes; the resulting voltage is measured across the inner electrodes (Fig. 3.13).

Assuming that only pure resistance is present, Ohm's law can be used to calculate the total impedance, $Z_T(t)$, as the ratio of voltage to current. The total thoracic impedance consists of a constant impedance, Z_0, and a time-varying impedance, $\Delta Z(t)$. Kubicek assumed that the change in thoracic impedance is related to the pulsatile volume change. He modeled constant tissue impedances such as bone, muscle, and fat as a conducting volume, in parallel with the pulsatile impedance. The empirical relationship he developed for one cardiac cycle, or beat, was:

$$SV(\text{beat}) = \rho \cdot \frac{L^2}{Z_0^2} \cdot LVET(\text{beat}) \cdot \frac{dZ}{dt_{MAX}}(\text{beat}), \tag{3.16}$$

where ρ is the resistivity of blood, L is the distance between inner band electrodes in cm, LVET(beat) is the left ventricular ejection time of a beat in seconds, and $dZ/dt_{MAX}(\text{beat})$ is the magnitude of the largest negative time derivative of the impedance change occurring

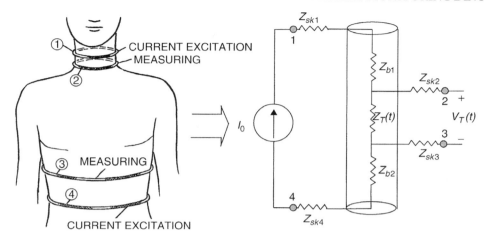

FIGURE 3.13: Kubicek's thoracic impedance parallel-column model and measurement setup. Based on (Kubicek et al. 1979)

during systole in ohms/s (Kubicek et al. 1979). As shown in Fig. 3.14, the impedance derivative is purposely inverted so that the original negative minimum change will appear as a positive maximum, in a manner more familiar to physicians.

Sramek chose to model the thorax as a truncated cone, rather than cylinder, and moved from band to spot electrode measurements. In collaboration with Bernstein, he modified the

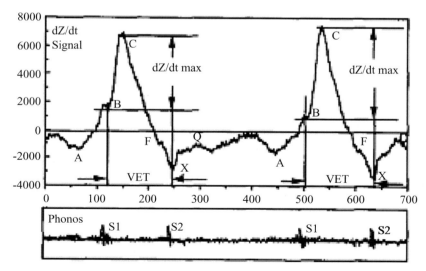

FIGURE 3.14: Idealized features of $-\mathrm{d}Z(t)/\mathrm{d}t$ signal, identifying fiducial points. From (Wang, 1995) © 2007 IEEE.

stroke volume equation to:

$$SV(beat) = \beta \cdot \left(\frac{Weight_{observed}}{Weight_{ideal}}\right) \left[\frac{(0.17H)^3}{4.2}\right] \cdot LVET(beat) \cdot \frac{\frac{dZ}{dt_{MAX}}}{Z_0}(beat), \qquad (3.17)$$

where β is the relative blood volume index, H is the patient height, $Weight_{observed}$ is the observed weight, and $Weight_{ideal}$ is the ideal weight (Sramek 1984, Bernstein 1986). Unfortunately, the correlation between these measurements from Sramek devices and those made using thermodilution was extremely low. In a study in which cardiac output was measured in 28 patients recovering from elective heart surgery, the square correlation coefficient between the two methods initially after surgery was $r^2 = 0.30 (p = 0.002)$, and decreased to $r^2 = 0.26$ ($p = 0.004$) two to four hours later (Yamikets and Jensen 1995).

3.3.2   Wavelet Transforms in CardioDynamics Software

Both Kubicek and Sramek calculated LVET(beat) and dZ/dt_{MAX}(beat) using heuristic, empirical methods of identification of curve changes and their time derivatives. As shown in Fig. 3.14, LVET(beat) is calculated from the B and X points, and dZ/dt_{MAX} (beat) is calculated from the C point. B is associated with aortic valve opening; X is associated with aortic valve closure. C represents a major upward deflection during systole. Each individual beat is parsed from R points in the QRS complex of a corresponding ECG waveform.

If dZ/dt waveforms were as clean as the idealized waveform in Fig. 3.14, heuristic detection could be sufficient. But since these waveforms, especially for critical cardiac patients, appear minimally periodic to the eye, accurate fiducial point detection is almost impossible with heuristic methods. As a first step to improving the correlation between impedance cardiography and thermodilution CO, Baura and Ng improved the detection of the fiducial points R, Q, C, B, and X. Using wavelet transforms of training data, noise due to respiration artifact and low signal-to-noise ratio was minimized, before detection. The training data were chosen from a data base of 266 patient waveforms. The effective sampling rate for these data, after downsampling, was 200 Hz (Baura and Ng 2003).

In general, R and Q point detection were based on scale 2 Haar detail coefficients, and enabled parsing of each ECG and dZ/dt beat. Within an impedance beat, C point detection was based on scale 1 Mallet approximation coefficients. The global maximum was determined in the search range of 1/3 and 1/2 of the length between the first and second Q points of dZ/dt. The first occurrence of these global maxima was designated the C point. B point detection was based on scale 2 Symlet 2 detail coefficients of ΔZ. Search for the most recent local maximum in the detail coefficients was limited between $[(R + 1)/4 + 1$ and $[(C - 2)/4 + 1)]$. The B point was detected as the global maximum in the range bracketed by this local maximum and

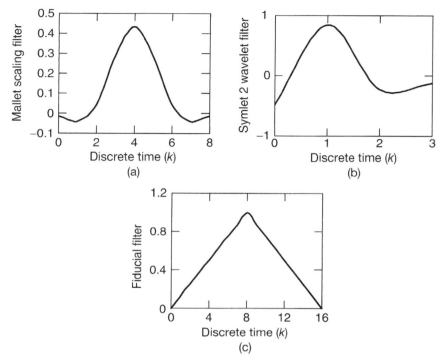

FIGURE 3.15: Scaling and wavelet filters used in fiducial point detection. (a) Mallet scaling filter, (b) Symlet 2 wavelet filter, and (c) Fiducial filter. Based on (Baura and Ng 2003)

(C − 2). X point detection was based on scale 1 approximation coefficients using a newly created Fiducial filter. Search for the X point was limited between the C point and the next Q point. The X point was detected as the first occurring local minimum of the scale 1 approximation coefficients (Baura 2002, Baura and Ng 2003). The Mallet scaling filter, Symlet 2 wavelet filter, and Fiducial filter are given in Fig. 3.15. As a representative example, B point detection is diagramed in Fig. 3.16. For more information on this detection system, please see (Baura 2002). Wavelet fiducial point detection is implemented in the CardioDynamics BioZ Dx monitor (Fig. 3.17).

An efficient process for research and development of these algorithms was recently described (Baura 2004b). In a large monitoring project, many algorithms may be invented in parallel. As shown in Fig. 3.18, the marketing department requests new monitoring features, which the research manager interprets as algorithm definitions.

Appropriate clinical data are collected, with a subset used for algorithm invention, or training, and the remainder reserved for algorithm validation, or testing. Algorithm validation is performed with predetermined validation criteria. If validation criteria are not met, the

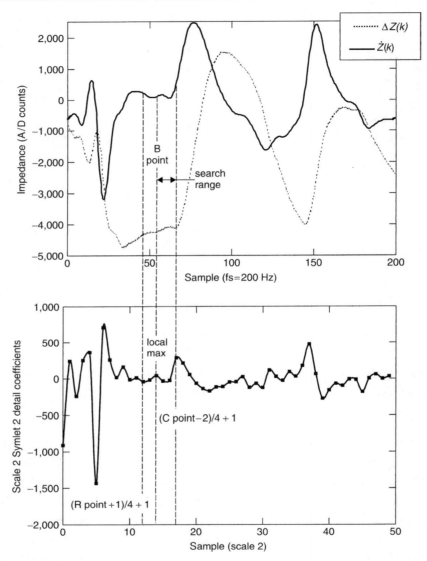

FIGURE 3.16: B point detection. This patient is a 62 years old, 82 kg pacemaker patient. The local minimum immediately after the Q point was caused by pacemaker stimulation. Based on (Baura and Ng 2003)

algorithms are retuned with training data. Validated algorithms are demonstrated to marketing, through a real-time implementation such as LabVIEW. Upon marketing approval, the LabVIEW algorithms are documented in specifications detailing the inputs, internal processing, and outputs of each algorithm module.

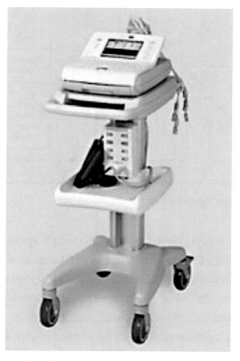

FIGURE 3.17: CardioDynamics BioZ Dx monitor. Courtesy of CardioDynamics, San Diego, CA

The transfer of algorithm specifications to software engineers initiates the algorithm development process (Fig. 3.19). Like other parts of a medical device development project, algorithm development is subject to the design control process, within the Good Manufacturing Practice requirements set forth by the Food and Drug Administration (FDA). The design input

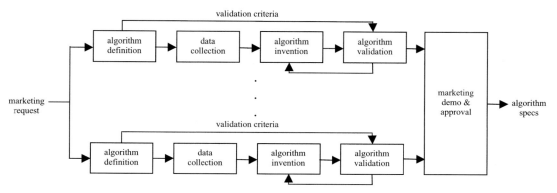

FIGURE 3.18: Algorithm research process

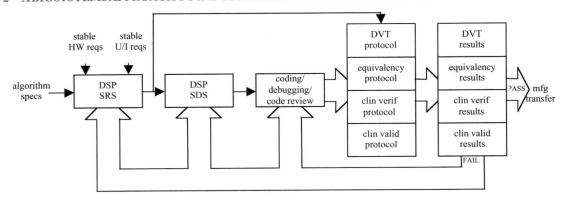

FIGURE 3.19: Algorithm development process

requirement is met by the digital signal processor (DSP) system requirements specification (SRS), which cannot be completed without stable hardware and user interface specifications. Note that the DSP SRS references the original LabVIEW algorithm specifications. The design output requirement is met by the DSP Software Design Specification (SDS), which provides the foundation for code creation, debugging, and code reviews.

Verification refers to confirmation that specific requirements have been met. This is proven through design verification testing (DVT) and clinical verification testing, per predefined protocols. The DVT protocol encompasses all tests required to prove that specific requirements in the SRS have been met. Clinical verification testing involves weekly testing at a clinical site, to verify that the algorithms respond as predicted to real patient data. Validation refers to confirmation that the particular requirements for a specific intended use can be consistently fulfilled. Using an equivalency protocol, testing is conducted to determine if the same parameter values are calculated for the same simulated inputs in the prototype device and a predicate device. Using a clinical validation protocol, testing is conducted in humans.

Results from verification and validation are fed back to update the SRS and SDS, as needed. As software bugs are exposed through testing, debugging occurs. When all tests have passed, the algorithms are transferred to manufacturing (Baura 2004b).

3.3.3   Clinical Results

Clinical data sets for validation were obtained from pacemaker patients at a local pacemaker clinic. In a single patient, stimulation at the baseline pacemaker stimulation amplitude setting, and when possible, at the minimum and maximum unipolar and bipolar amplitudes, results in up to five very different ECG and dZ/dt waveforms. In total, 63 waveforms from 28 patients were collected for validation (Baura 2004a). The effective sampling rate for these data, after downsampling, was 200 Hz. Three beats each from 58 validation data sets were used for fiducial point detection validation. The validation criteria were based on practical clinical needs in the field. The mean absolute error (MAE) was used for validation criteria, as it is more robust than mean error, which enables positive and negative errors to cancel.

For R point detection, the validation criterion was MAE $\leq$ 2 samples. The R point validation error was MAE = 0 samples. For Q point detection, the validation criterion was MAE $\leq$ 5 samples. The Q point validation error was MAE = 3 samples. For C point detection, the validation criterion was MAE $\leq$ 2 samples. The C point validation error was MAE = 1 sample. Finally, for B and X point detection, the validation criterion was LVET MAE $\leq$ 5 samples. The LVET MAE was 4 samples (Baura 2002, 2004a, Baura and Ng 2003).

3.3.4   Conclusion

When the Connors et al. study was published in 1996, the market for invasive CO measurements began to decline. However, some predicted that this study would increase the market for noninvasive cardiac output measurements. This did not occur instead, it decreased the market for all cardiac output measurements. The decrease in thermodilution catheter sales from 1997 to 2006 was detailed earlier in this chapter. Similarly, nonfederal hospital sales of impedance cardiography monitors from 2000, 2003, and 2006 generated $x,xxx,xxxx, $x,xxx,xxx, and $x,xxx,xxx, respectively, in the United States (IMS Health 2000, 2003, 2006).

3.4   ASPECT MEDICAL DEPTH OF ANESTHESIA MONITORING

Aspect Medical was founded in 1987 by electrical engineer Nassib Chamoun, with the intent of directly monitoring depth of anesthesia. Anesthesia can be defined as a state of drug-induced unconsciousness, in which a patient neither perceives nor recalls noxious stimulation (Prys-Roberts 1987). Historically, this continuum from minimal sedation to general anesthesia was monitoring by an anesthesiologist during surgery by observing responsiveness to verbal or tactile stimulation, spontaneous ventilation, and cardiovascular function.

However, some patients postoperatively report awareness, that is, the postoperative recollection of events occurring during general anesthesia. Affected patients report perception of paralysis, conversations, and surgical manipulations, accompanied by feelings of helplessness, fear, and pain. The incidence of awareness is 0.1–0.2% in the general surgical population, but is greater during cardiac surgery, caesarian section, and trauma surgery (Myles et al. 2004). In the 1990 timeframe, depth of anesthesia monitoring was considered by venture capitalists to be one of the patient monitoring "holy grails". Aspect Medical believed that bispectrum analysis of the electroencephalogram (EEG) could be used to monitor depth of anesthesia.

3.4.1   Depth of Anesthesia Monitoring

Clinical signs such as blood pressure and heart rate are routinely used to monitor depth of anesthesia, but may be unreliable. Alternatively, methods based on EEG processing have been investigated. In 1937, Gibbs et al. first reported that the influence of certain drugs on nervous activity could be observed in the EEG (Gibbs et al. 1937). Later, Falconer and Bickford noted that the electrical power in EEG was associated with changes in the rate of thiopental or ethyl ether administration (Bickford 1950). However, power is sensitive to electrode location, and insensitive to important changes in frequency distribution.

In the frequency domain, various parameters were created to explore potential correlation with depth of anesthesia. Two such parameters are the peak power frequency and spectral edge frequency (highest EEG frequency). A third parameter is the bispectrum, $B(f_1, f_2)$:

$$B(f_1, f_2) = \left| \sum_{i=1}^{L} X_i(f_1) X_i(f_2) X_i^*(f_1 + f_2) \right|, \qquad (3.18)$$

where i is the epoch number, L is the total number of epochs summed, f is a selected frequency, $X(f)$ is the Fourier transform of $x(k)$, and $X^*(f)$ is the complex conjugate of $X(f)$. None of these parameters singly has been demonstrated to correlate well with depth of anesthesia (Rampil 1998).

3.4.2   System Identification in Aspect Medical Software

Aspect Medical used a true black box approach to develop a parameter that correlates with depth of anesthesia. First, approximately 5000 h of EEG were recorded, representing about 1500 anesthetic administrations and a variety of anesthetic protocols. The awareness behavior of each patient during the course of each recording was noted. A range of prospective subparameters was calculated, and each correlation with awareness behavior was tested. The parameters with the best performance were entered into a multivariate analysis for the creation of a final parameter called the bispectral index, BIS (Chamoun et al. 1995).

The first subparameter, burst suppression ratio (BSR), is the fraction of epoch length where EEG voltage is suppressed (does not exceed ±5 mV). The QUAZI suppression index is designed to detect burst suppression, in the presence of wandering baseline voltage. BetaRatio is the log ratio of power in two empirically derived frequency bands: 30–47 and 11–20 Hz. SynchFastSlow is the log of the ratio of the sum of all bispectrum peaks in the area from 0.5 to 47 Hz, over the sum of the bispectrum in the area 40–47 Hz. When all four subparameters are combined as BIS, BetaRatio is weighted most heavily when EEG has the characteristic of light sedation. SynchFastSlow predominates during the phenomenon of EEG activation

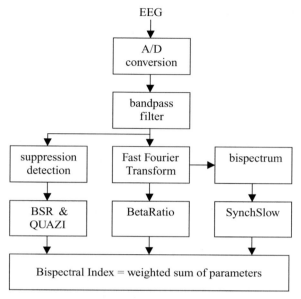

FIGURE 3.20: Flowchart of BIS calculations. Based on (Rampil 1998)

and during surgical level of hypnosis. BSR and QUAZI detect deep anesthesia (Rampil 1998).

A flowchart for calculation of BIS is given in Fig. 3.20. The BIS index range of 0–100 correlates well with flatline EEG through general anesthesia and the awake state (Fig. 3.21). Aspect Medical's BIS Vista standalone monitor is given in Fig. 3.22.

3.4.3 Clinical Results

The B-Aware trial investigated the ability of BIS to prevent awareness during anesthesia administration. In this double-blinded, randomized study, 1225 surgical patients were assigned to the BIS group and 1238 patients were assigned to the routine care group. All patients received a BIS sensor on the forehead before anesthesia induction, but the BIS monitor was not powered on for routine care patients. For the BIS group, the anesthesiologist adjusted anesthesia delivery to maintain a BIS value of 40–60, from the start of laryngoscopy to the time of wound closure. For the BIS group of patients, BIS was manually recorded at 5 min intervals for the first hour, and every 10 min thereafter.

After surgery, each patient was interviewed three times to determine if awareness had occurred, since postoperative recollections can be transient. The number of awareness reports for the BIS and routine care groups was 2 and 11 ($p = 0.022$), respectively, a significant awareness risk reduction of 82% (Myles et al. 2004).

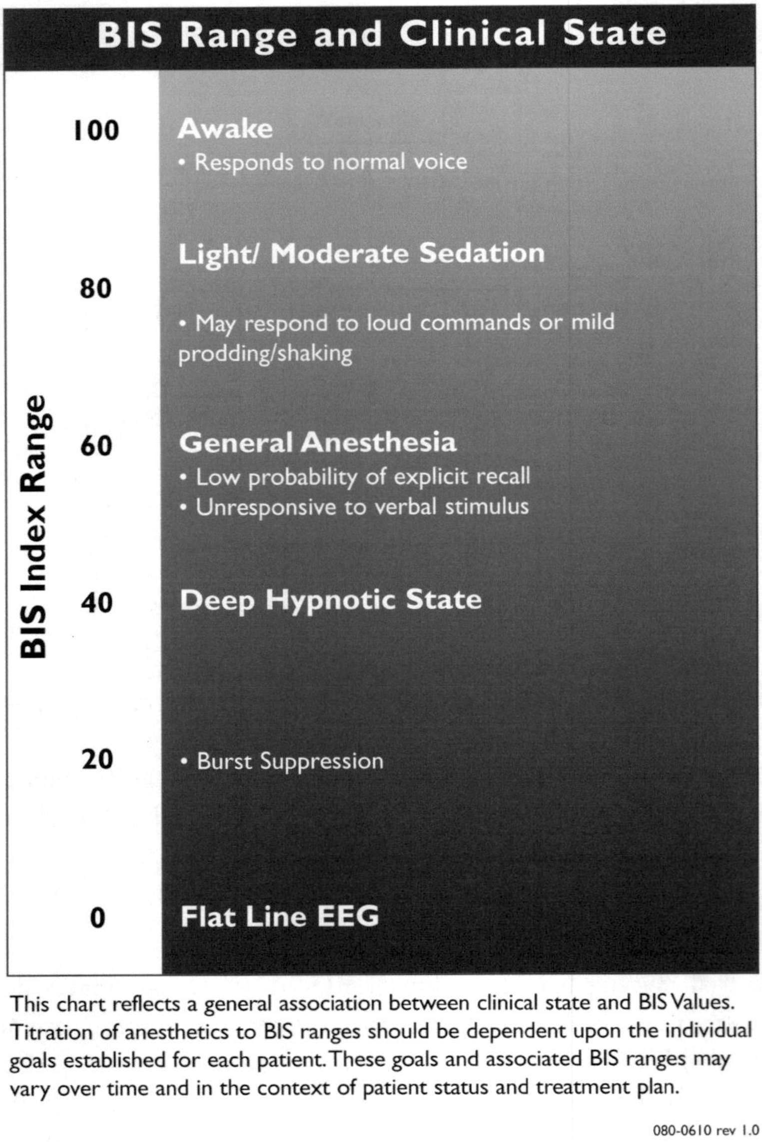
BIS Range and Clinical State
BIS Index Range
100
Awake
• Responds to normal voice
Light/ Moderate Sedation
80
• May respond to loud commands or mild prodding/shaking
60
General Anesthesia
• Low probability of explicit recall
• Unresponsive to verbal stimulus
40
Deep Hypnotic State
20
• Burst Suppression
0
Flat Line EEG
This chart reflects a general association between clinical state and BIS Values. Titration of anesthetics to BIS ranges should be dependent upon the individual goals established for each patient. These goals and associated BIS ranges may vary over time and in the context of patient status and treatment plan.
080-0610 rev 1.0

FIGURE 3.21: The relationship between BIS and depth of anesthesia. Courtesy of Aspect Medical Systems, Norwood, MA

In a similar study named SAFE-2, 4945 surgical patients were monitored with BIS, and compared to 7836 historical patients receiving routine care. The number of awareness reports for the BIS and routine care groups was 2 and 14 ($p < 0.038$), respectively, a significant awareness risk reduction of 86% (Ekman et al. 2004).

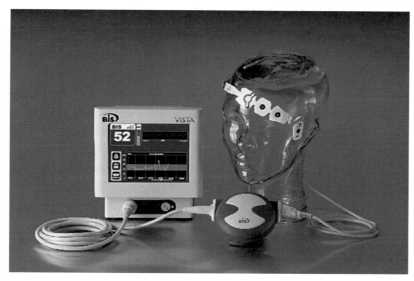

FIGURE 3.22: Aspect Medical BIS Vista standalone monitor. Courtesy of Aspect Medical Systems, Norwood, MA

3.4.4   Conclusion

In October 2003, FDA granted Aspect Medical clearance to indicate BIS for reducing the risk of intraoperative awareness. No other device possesses this FDA intraoperative awareness reduction indication. In late 2004, after 18 years and $250 million in funding, Aspect Medical reached its first quarterly profit. In recent years, Aspect Medical has experienced a growth rate of around 20% and a profit margin around 75% (Swain 2004).

In 2000, nonfederal hospital sales of BIS monitors and their monitoring accessories generated $xx,xxx,xxx and $xx,xxx,xxx, respectively (IMS Health 2000). By 2006, nonfederal hospital sales had increased to $xx,xxx,xxx and $xx,xxx,xxx, respectively (IMS Health 2006). Aspect Medical is in the process of increasing bispectral index utility through diagnostic investigations of other brain applications. BIS may be predictive of Alzheimer's disease and antidepressive medication effectiveness.

REFERENCES

American Society of Anesthesiologists, Standards for postanesthesia care, *ASA Directory of Members*, 56th ed., 1991, p. 672.

Barker, S. J. and Shah, N. K., "The effects of motion on the performance of pulse oximeters in volunteers (revised publication)," *Anesthesiology*, Vol. 86, pp. 101–108, 1997.

Baura, G. D., *System Theory and Practical Application of Biomedical Signals*. Hoboken, NJ: Wiley-IEEE Press, 2002.

Baura, G. D. and Ng, S. K., "Method and apparatus for hemodynamic assessment including fiducial point detection," U.S. Patent 6,561,986, May 13, 2003.

Baura, G. D., "Listen to your data," *IEEE Signal Proc Mag*, Vol. 21, pp. 21–25, 2004a.

—— "System theory in industrial patient monitoring: an overview," in *Proc. of 26th Annual Int. Conf. of IEEE EMBS*, San Francisco, Sept. 1–5, 2004b, pp. 5356–5359.

Bernstein, D. P., "A new stroke volume equation for thoracic electrical bioimpedance: theory and rationale," *Crit Care Med*, Vol. 14, pp. 904–909, 1986.

Bickford, R. G., "Automatic electroencephalographic control of general anesthesia," *Electroencephalogr Clin Neurophysiol*, Vol. 2, pp. 93–96, 1950.

Chamoun, N. G., Sigl, J. C., and Smith, C. P., "Cerebral biopotential analysis system and method," U.S. Patent 5,458,117, October 17, 1995.

Connors, A. F., Speroff, T., Dawnson, N. V., Thomas, C., Harrell, F. E., Wagner, D., Desbiens, N., Goldman, L., Wu, A. W., Califf, R. M., Fulkerson, W. J., Vidaillet, H., Broste, S., Bellamy, P., Lynn, J., and Knaus, W. A., "The effectiveness of right heart catheterization in the initial care of critically ill patients," *JAMA*, Vol. 276, pp. 889–897, 1996.

Diab, M. K., Kiani-Azarbayjan, E., Elfadel, I. M., McCarthy, R. J., Weber, W. M., and Smith, R. A., "Signal processing apparatus," U.S. Patent 5,632,272, May 27, 1997.

Ekman, A., Lindholm, M.-L., Lennmarken, C., and Sandin, R., "Reduction in the incidence of awareness using BIS monitoring," *Acta Anaesthesiol Scand*, Vol. 48, pp. 20–26, 2004.

Gibbs, F. A., Gibbs, E. L., and Lennox, W. G., "Effect on the electroencephalogram of certain drugs which influence nervous activity," *Arch Intern Med*, Vol. 60, pp. 154–166, 1937.

IMS Health, *Hospital Supply Index*. Plymouth Meeting, PA: IMS America, 1997. IMS Health, Hospital Supply Index. Plymouth Meeting, PA: IMS America, 2000.

IMS Health, *Hospital Supply Index*. Plymouth Meeting, PA: IMS America, 2003.

IMS Health, *Hospital Supply Index*. Plymouth Meeting, PA: IMS America, 2006.

Kubicek, W. G., Kinnen, E., Patterson, R. P., and Witsoe, D. A., "Impedance plethysmograph," U.S. Patent Re. 30,101, September 25, 1979.

Mackenzie, N., "Comparison of a pulse oximeter with an ear oximeter and an in vitro oximeter," *J Clin Monit*, Vol. 1, pp. 156–160, 1985.

Masimo Corporation, *Patent Litigation Status*, http://www.masimo.com/patentLitigation.htm. December 2006.

Myles, P. S., Leslie, K., McNeil, J., Forbes, A., and Chan, M. T. V., "Bispectral index monitoring to prevent awarenss during anaesthsia: the B-Aware randomized controlled trial," *Lancet*, Vol. 363, pp. 1757–1763, 2004.

Prys-Roberts, C., "Anaesthesia: a practical or impractical construct?," *Br. J. Anaesth.*, Vol. 59, pp. 1341–1345, 1987.

Rampil, I., "A primer for EEG signal processing in anesthesia," *Anesthesiology*, Vol. 89, pp. 980–1002, 1998.

Sramek, B., "Noninvasive continuous cardiac output monitor," U.S. Patent 4,450,527, May 22, 1984.

Swain, E., "Patience pays off," *MDDI*, October, pp. 18–24, 2004.

Wang, X., Sun, H. H., and Van De Water, J. M., "An advanced signal processing technique for impedance cardiography," *IEEE Trans Biomed*, Vol. 42, pp. 224–230, 1995.

Yamikets, J. and Jensen, L., "Evaluation of impedance cardiography: comparison of NC-COM3-R7 with Fick and thermodilution methods," *Heart & Lung*, Vol. 24, pp. 194–206, 1995.

Yelderman, M. L., "Method and apparatus for measuring flow," U.S. Patent 4,507,974, April 2, 1985.

—— "Continuous measurement of cardiac output with the use of stochastic system identification techniques," *J. Clin. Monit.*, Vol. 6, pp. 322–332, 1990.

Yelderman, M. L., Quinn, M. D., McKown, R. C., Eberhart, R. C., and Dollar, M. L., "Continuous thermodilution cardiac output measurement in sheep," *J. Thorac. Cardiovasc. Surg.*, Vol. 104, pp. 315–320, 1992a.

Yelderman, M. L., Ramsay, M. A., Quinn, M. D., Paulsen, A. W., McKown, R. C., and Gillman, P. H., "Continuous thermodilution cardiac output measurement in intensive care unit patients," *J. Cardiothorac. Vasc. Anesth.*, Vol. 6, pp. 270–274, 1992b.

CHAPTER 4

Diagnostic Devices

In this chapter, we describe a diagnostic device that uses system theory for classification. Since medical diagnostics involves discrete testing, it is imperative that this one-shot test result be accurate.

Neopath used fuzzy models to classify Pap smear cervical cells. While other diagnostic devices in development use system theory, none has achieved widespread use and high clinical accuracy.

4.1 NEOPATH CERVICAL CANCER SCREENING

Neopath was founded in 1989 by bioengineering professor Dr. Alan Nelson, with the intent of automating cervical cancer screening. Cervical cancer is one of the most common malignancies in women, accounting for 15,700 new cases and 4900 deaths in the United States each year. The standard diagnostic test for cervical cancer screening is the Papanicolaou (Pap) smear, during which cervical cells are collected during a pelvic exam and preserved on a glass slide for classification. Since its introduction in the United States in 1947, the Pap smear has been credited with decreasing the incidence of cervical cancer from 44 per 100,000 to 5–8 per 100,000 women. However, Pap smear false-negatives do occur, with about 67% resulting from sampling error (abnormal cells not placed on smear) and 33% resulting from detection error during the pre-1995 timeframe (Agency for Healthcare Research and Quality 2000). Detection is performed by certified cytotechnologists, who view Pap smears under a microscope and classify abnormal cells.

During the late 1980s and early 1990s, several technologies were being developed to reduce this false-negative rate. Thin-layer cytology aimed to reduce the sampling error by transferring cervical cell samples to slides with fewer artifacts. Neopath and competitor Neuromedical Systems aimed to reduce the detection error through computer rescreening technologies.

4.1.1 Pap Smear Screening

Classification of cervical cell samples was first introduced as a cervical cancer detection method by Babes in 1926. During the same time frame, George Papanicolaou noted that abnormal cells were present in the vaginal pool in the presence of early cervical cancer. In 1943, Papanicolaou and Traut published the first book on the diagnosis of cervical neoplasia (new cell growth) by vaginal pool smear. Later, Papanicolaou, with other investigators, developed a classification system based on the degree of abnormality of the cells on the smear. The Pap smear classification system was revised in 1988 by the National Cancer Institute in Bethesda, Maryland, based on the work (Patten 1978) of noted cytopathologist Dr. Stanley Patten, Jr. (Bonfiglio 1997).

The Bethesda System (TBS) classifies cervical cells and evaluates the specimen for quality. The two main cellular diagnoses are benign cellular changes and epithelial abnormalities. Classification of epithelial abnormalities is based on the progression of cervical dysplasia. Normally, the cervical lining is composed of organized layers of uniformly shaped cells, with the bottom layer containing round cells. As these cells mature, they rise to the surface and flatten out to become flat squamous cells. During mild cervical dysplasia, which is designated as CIN I, this growth process is disrupted with a few abnormal cells. During moderate dysplasia, or CIN II, the abnormal cells are distributed in about half the thickness of the lining. If the abnormal growth processes to severe dysplasia or carcinoma-in-situ (known as CIN III), the entire thickness becomes disordered, but the abnormal cells have not spread below the lining. Once the abnormal growth invades the tissue, it becomes invasive cancer (Fig. 4.1).

Epithelial cell abnormalities can be classified into six categories. Atypical squamous cells of undetermined significance (ASCUS) classification refers to unusual cells that are not abnormal enough, compared to the benign condition, to be classified as dysplasia. This classification usually results in a second Pap smear for verification. Low-grade squamous intraepithelial lesion (LSIL) classification is associated with hollow cells that possess atypical nuclei and/or mild dysplasia. Detection of LSIL cells usually results in colposcopy (visual examination using a lighted magnifying instrument) for further examination. High-grade squamous intraepithelial lesion (HSIL) classification is associated with moderate or severe dysplasia or carcinoma-in-situ. The cells are usually deficient in the dense region of the nucleus containing DNA, are granular or reticular, and are often found in patterns of lines (Fig. 4.2).

Squamous cell carcinoma classification refers to a malignant invasive tumor of squamous cells. These cells occur singly or in aggregates; their nuclei contain coarse granular clumps (Fig. 4.3). Adenocarcinoma classification refers to a malignant invasive tumor composed of endocervical, endometrial (from the lining of the uterus), or extrauterine cells. Finally, atypical glandular cells of undetermined significance (AGUS) classification is associated with cellular changes in glandular cells exceeding those expected in a benign reactive or reparative reaction. Detection of AGUS cells usually results in colposcopy, as well as scraping of the inner uterine lining.

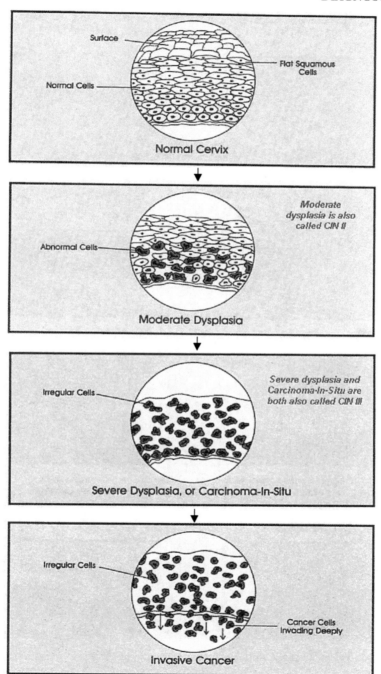

FIGURE 4.1: Progression of cervical dysplasia from normal to invasive cancer. Courtesy of Paul Indman, MD, www.gynalternatives.com, Los Gatos, CA

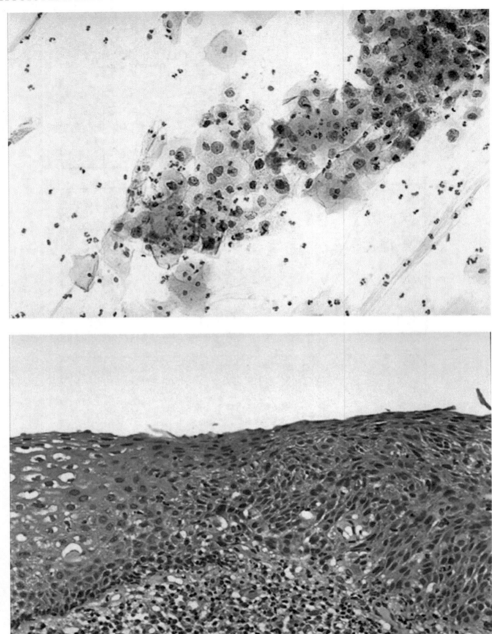

FIGURE 4.2: (a) HSIL Pap smear and (b) cervical biopsy. The biopsy contains normal cells on the left and dysplasic cells on the right. The smear and biopsy were obtained from different patients. Images from WebPath (http://library.med.utah.edu/WebPath/webpath.html), courtesy of Edward C. Klatt, MD

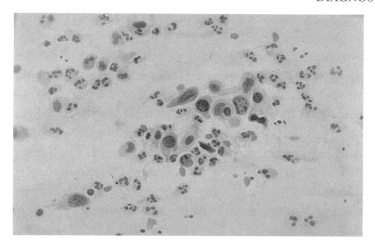

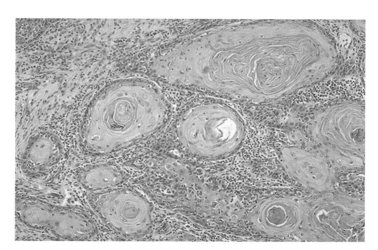

FIGURE 4.3: (a) Squamous cell carcinoma Pap smear and (b) cervical biopsy. The smear and biopsy were obtained from different patients. Images from WebPath (http://library.med.utah.edu/WebPath/webpath.html), courtesy of Edward C. Klatt, MD

4.1.2   Fuzzy Models in Neopath Software

Neopath pioneered cervical cell classification using fuzzy models. Likely, they used fuzzy models because the application of artificial neural networks to cell classification had already been patented by competitor Neuromedical Systems in 1990 (Rutenberg, 1990). Through sheer luck, Dr. Stanley Patten, Jr., whose work laid the foundation for The Bethesda System classification, happened to retire in Seattle, near where Neopath was based. Patten was lured out of retirement to become Neopath's Medical Director. There, he taught Neopath engineers about cervical cytology, and once he learned the theory underlying fuzzy models, converted his personal cytology rules to this format.

Pattens' discrimination rules for different cell types, which are implemented in the Neopath AutoPap system, are illustrated in Table 4.1. Each rule should be interpreted as the intersection of six conditions to describe a particular cell type. The rules were developed from a training set of 4174 slides (Lee and Nelson 1997). The system configuration for rescreening is given in Fig. 4.4. A Becton Dickinson Focal Point System, which is a direct descendent of the Neopath AutoPap Primary Screening System, is given in Fig. 4.5.

4.1.3 Clinical Results

Both the Neopath and Neuromedical Systems were approved by the Food and Drug Administration (FDA) for rescreening of Pap smears in late 1995. However, only the Neopath system was approved for primary screening (cytotechnologist substitute) in 1998. As of 2007, the Neopath system is still the only system approved for primary screening of Pap smears. The results of the clinical trial for primary screening are given below.

The primary screening trial involved 25,124 analyzed slides from five commercial laboratories. Each slide was screened by one cytotechnologist and by the Neopath AutoPap system. If each slide classification pair was not in agreement, a panel of two to three cytotechnologists reviewed the slide in question. If no agreement could be reached, then a panel of three cytopathologists reviewed the slide for a consensus truth determination. The sensitivities for the AutoPap and cytotechnologists were calculated for the difficult classes ASCUS+, LSIL, LSIL+, and HSIL (Table 4.2).

For all classes, the AutoPap had higher sensitivity. Further, these sensitivities were significantly different ($p < 0.013$) for the LSIL, LSIL+, and HSIL classes (Wilbur et al. 1999).

4.1.4 Conclusion

In 1999, AutoPap merged with AutoCyte, a thin-layer cytology manufacturer, and acquired the intellectual property of Neuromedical Systems. The merged company was renamed TriPath Imaging. A combined AutoCyte/AutoPap system obtained FDA approval in 2001. On December 20, 2006, TriPath Imaging was acquired by Becton Dickinson.

Automation has changed cytotechnology practice. Computer-assisted screening and primary screening have increased sensitivity, productivity, and consistency, while decreasing turnaround time. And the next change is on the horizon. The human papillomavirus (HPV), which is the most commonly sexually transmitted infection in the U.S., is linked to cervical cancer. In fact, four HPV types are responsible for 70% of cervical cancers. On June 8, 2006, FDA approved Merck's Gardasil HPV vaccine for the prevention of cervical cancer.

TABLE 4.1: Neopath discrimination rules for different cervical cell types. Reprinted from (Lee and Nelson 1997) with permission from John Wiley & Sons

	CHROMATIN PARTICLES	CHROMATIN DISTRIBUTION	CELL SHAPE	CYTO TEXTURE	CELL BORDERS	CELL ARRANGEMENT
Intermediate squamous	Fine	Even	Polygonal	Homogeneous	T	Isolated
Endometrial	Fine	Even	Round	Finely vacuolated	F	Isolated/cluster
Squamous metaplastic (P)	Fine	Even	Round/oval	Homogeneous	F	Isolated
Squamous metaplastic (I)	Fine	Even	Round/oval	Primitive	T	Isolated
Squamous metaplastic (M)	Fine	Even	Small polygonal	Homogeneous	T	Isolated
Atypical endometrial hyperplasia	Fine	Even	Round	Vacuolated	F	Cluster
Endometrial Adenoca	Fine	Uneven	Round	Vacuolated	F	Isolated/cluster
Endocervical	Fine	Even	Columnar	Granular	T	Isolated/sheets
Atrophic squamous	Fine	Even	Round/oval	Homogeneous	T	Isolated
Squamous atypical repair	Fine	Even	Round/oval	Homogeneous	T	Sheets
ASCUS	Fine/oval	Even	Polyg/round	Homogeneous	T	Isolated
Metaplastic dysplasia	Fine/clumped	Even	Round/oval	Homogeneous	T	Isolated
Intermediate CIS	Fine/coarse	Even	None	Primitive	F	Isolated/syncytial
Large cell CIS	Fine	Even	None	Primitive	T	Isolated/syncytial
Nonkeratinizing sq CA	Fine/coarse	Uneven	None	Primitive	F	Isolated/syncytial
Keratinizing dysplasia	Opaque	Even	Pleomorphic	Homogeneous	T	Isolated
Keratinizing sq carcinoma	Opaque	Even/uneven	Pleomorphic	Homogeneous	T	Isolated
Small cell CIS	Fine/coarse	Even	Oval	Primitive	F	Syncytial
Small cell sq carcinoma	Fine/coarse	Uneven	Oval	Primitive	F	Syncytial
Atypical immature squamous metaplastic type	Fine	Even	Round	Homogeneous	T	Isolated
Endocervical atypia	Fine	Even	Columnar/oval	Homogeneous	T	Sheets
Adenocarcinoma in situ	Coarse	Even	Columnar	Granular	T	Isolated/sheets
Endocarvical adenoca	Fine	Uneven	Columnar	Granular	F	Cluster

P: primitive, I: immature, M: mature, Adenoca: adenocarcinoma, ASCUS: atypical squamous cells of undetermined significance, CIS: carcinoma in situ, sq: squamous, CA: carcinoma, T: true, F: false.

TABLE 4.2: Sensitivities for the AutoPap System and Cytotechnologists. Based on (Wilbur et al. 1999)

	AUTOPAP (%)	CYTOTECHNOLOGISTS (%)
ASCUS+	86	79
LSIL	91	84
LSIL+	92	86
HSIL+	97	93

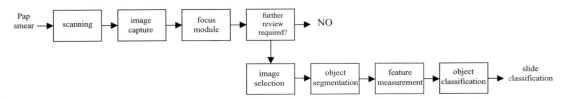

FIGURE 4.4: Neopath AutoPap rescreening system configuration. Based on (Neopath 1995)

FIGURE 4.5: Beckton Dickinson Focal Point System. Courtesy and © Becton, Dickinson and Company.

As more women receive the HPV vaccine, it is expected that the rate of cervical cancer will decline sharply (Centers for Disease Control and Prevention 2006). The Pap smear, a mainstay against cervical cancer, will become more costly per cancer case detected, and could decline in use.

REFERENCES

Agency for Healthcare Research and Quality, *Evaluation of Cervical Cytology.* http://www.ahrq.gov/clinic/epcsums/cervsumm.htm. November 2000.

Bonfiglio, T. A., "Diagnostic cytology of the uterine cervix: a major contribution and classic reference in gynecologic cytopathology," *Cancer Cytopath*, Vol. 81, pp. 324–327, 1997.

Centers for Disease Control and Prevention, *Human Papillomavirus: HPV information for clinicians.* Atlanta, GA: CDC, 2006.

Lee, J. S. L. and Nelson, A. C., "Stanley F. Patten, Jr., M.D., Ph.D., and the development of an automated Papanicolaou smear screening system," *Cancer Cytopath*, Vol. 81, pp. 332–336, 1997.

Neopath, Premarket Approval of the AutoPap 300 QC System. *P950009*, September 29, 1995.

Patten, S. F., Jr., *Diagnostic Cytopathology of the Uterine Cervix.* 2nd ed. Basel: S Karger, 1978. Rutenburg, M. R., "Neural network based automated cytological specimen classification system and method," U.S. Patent 4,965,725, October 23, 1990.

Wilbur, D. C., Prey, M. U., Miller, W. M., Pawlick, G. F., Colgan, T. J., and Taylor, D. D., "Detection of high grade squamous intraepithelial lesions and tumors using the AutoPap system," *Cancer Cytopath.*, Vol. 87, pp. 354–358, 1999.

Conclusion

In this Lecture, we described some of the common system theory techniques that are part of the toolkit of medical device engineers in industry. These techniques include the pseudorandom binary sequence, adaptive filtering, wavelet transforms, the autoregressive moving average model with exogenous input, artificial neural networks, fuzzy models, and fuzzy control. Because the clinical usage requirements for patient monitoring and diagnostic devices are so high, system theory is the preferred substitute for heuristic, empirical processing during noise artifact minimization and classification.

We also discussed some exemplary applications of system theory processing that have been deployed. Masimo used adaptive filtering to minimize motion artifact during pulse oximetry measurements. Interflo Medical used the pseudorandom binary sequence to increase the signal-to-noise ratio during continuous thermodilution measurements. CardioDynamics used wavelet transforms to minimize the noise artifact during impedance cardiography measurements. Aspect Medical Systems used black box system identification to identify the stages of anesthesia administration. Neopath used fuzzy models to classify Pap smear cervical cells. As new applications debut, they will be added to this Lecture.

Author Biography

Gail Dawn Baura received a BSEE from Loyola Marymount University in 1984, and an MSEE and MSBME from Drexel University in 1987. She received a PhD in Bioengineering from the University of Washington in 1993. Between these graduate degrees, Dr. Baura worked as a loop transmission systems engineer at AT&T Bell Laboratories. Since graduation, she has served in a variety of research and development positions at IVAC Corporation, Cardiotronics Systems, Alaris Medical Systems, and VitalWave Corporation (now Tensys Medical). Her most recent industrial position was Vice President of Research and Chief Scientist at CardioDynamics. In 2006, she returned to academia as a Professor at Keck Graduate Institute of Applied Life Sciences in Claremont, CA.

Dr. Baura's textbook, System Theory and Practical Application of Biomedical Signals (Wiley-IEEE Press, 2002), is part of the IEEE Series in Biomedical Engineering. Her textbook, Engineering Ethics: An Industrial Perspective (Academic Press, 2006), was written to provide industrial material for engineering programs wishing to fulfill the accreditation ethics requirement. Dr. Baura is a senior member of IEEE, associate editor of IEEE Engineering in Medicine and Biology Magazine, and a biomedical engineering evaluator for the Accreditation Board of Engineering Technology. She holds 15 issued and 6 pending U.S. patents. Her research interests are the application of system theory to patient monitoring and other devices.

Printed in the United States
by Baker & Taylor Publisher Services